Learning Statistics and Excel in Tandem

for Excel 97

by

Saroja Selvanathan
School of Economics
Faculty of International Business and Politics
Griffith University
Queensland, Australia

and

E.A. Selvanathan
School of Economics
Faculty of International Business and Politics
Griffith University
Queensland, Australia

THOMSON

Australia · Canada · Mexico · Singapore · Spain · United Kingdom · United States

To our parents:

Mr Murugesapillai Pathmanathan
Mrs Parameswary Pathmanathan
(late) Mr Velupillai Eliyathamby
Mrs Pooranam Eliyathamby

THOMSON

Level 7, 80 Dorcas Street
South Melbourne Victoria 3205

Email highereducation@thomsonlearning.com.au
Website http://www.thomsonlearning.com.au

First published in 2000
12 11 10 9 8
10 09 08 07 06

National Library of Australia
Cataloguing-in-Publication data

ISBN 0 17 010237 8

Printed in Australia by Ligare Book Printers
This title is printed under the imprint of Thomson.
Nelson Australia Pty Limited ACN 058 280 149 (incorporated in Victoria)
trading as Thomson Learning.

Preface

In most practical business applications, statistical data analysis involves dealing with large amounts of data. Universities, business and industrial organisations use at least one spreadsheet software package for statistical data analysis. Microsoft EXCEL is the most popular and the most commonly used spreadsheet package in businesses around the world. Most universities have incorporated Microsoft EXCEL into their curriculum.

Learning Statistics and EXCEL in Tandem shows how to solve most statistical business problems taught in an introductory statistics subject using the Microsoft EXCEL spreadsheet software. The major difference between this book and other EXCEL books is that, before using EXCEL for an application, in this book, we discuss the selection of the appropriate statistical technique required to solve the problem. In other words, the aim of the book is to teach in tandem the statistical techniques as well as the use of EXCEL to execute them. In this sense, this book may be used as a stand-alone book by managers and analysts who have knowledge in statistics. However, those with no statistics knowledge should use this as a supplement to a statistics textbook.

Learning Statistics and EXCEL in Tandem is primarily designed to cater for undergraduate students enrolled in a statistics subject. The book is also useful for postgraduate students as well as managers and analysts in analysing their day-to-day basic statistical problems using EXCEL.

Each chapter clearly demonstrates a specific topic in statistics. In each chapter, we clearly state the aim of the chapter, the statistical theory and techniques used in the chapter, and then show, using worked examples, how a real world problem related to that topic can be solved using EXCEL. Each example is worked out step-by-step showing the appropriate statistical technique and the EXCEL instructions required. Furthermore, at the end of each chapter, a number of similar exercises are given for the user to practice and test his/her understanding of the topic and the use of EXCEL commands related to that topic.

The most recent versions of EXCEL are EXCEL 5.0 for Windows 3.1 (or EXCEL 5), EXCEL 7.0 for Windows 95 (or EXCEL 7), and Microsoft Excel97. While these three versions are similar, there are some differences. Most of the EXCEL instructions in this book are based on Microsoft Excel97. However, if you have EXCEL 5 or EXCEL 7.0, you can still use most parts of the book.

We would like to thank the college editor, Glen Sheldon, ITP Nelson, Australia, for his encouragement throughout the duration of the project. Also our sincere thanks to Professor Prasada Rao, University of New England; Andrew Paltridge, Queensland University of Technology; Viji Krishnaratnam, Jen Riley and Gopi Gnananthan, Griffith University; and a number of anonymous reviewers for their valuable comments on various chapters of the manuscript during the development stages which has benefited this book. Last but not least, a special thank you to our children, Arthavan Selvanathan (7 years) and Prabha Selvanathan (5 years) for their love, patience and understanding throughout the duration of the project.

We would like to acknowledge the financial support from the School of Information Systems and Management Science, Griffith University, for this project.

We welcome comments about the book and suggestions for improvements.

Saroja Selvanathan
Antony Selvanathan

Table of Contents

PART THREE STATISTICAL INFERENCE

***A note to teachers:** Chapters 14-17 listed below are not included in this edition. However, they are available upon request from the authors if required in your subject.*

Table of Figures

Chapter 1

Introduction to Microsoft EXCEL

1.1 Accessing Microsoft EXCEL

When you start the computer the **EXCEL** icon will appear on the main Menu. Click the mouse on the **EXCEL** icon to activate EXCEL. If the **EXCEL** icon does not appear straight away, click on 'Start', select Programs and click on **Microsoft Excel**.

1.2 Introduction to EXCEL

You are now in EXCEL and what you now see on the screen is an **EXCEL Workbook Window** with a **Worksheet** (also known as a **Spreadsheet**) shown in Figure 1.1. The workbook is made up of a number of worksheets.

The EXCEL workbook window you see on the screen may sometimes fill only a portion of your screen. To maximise the workbook area (i.e., to fill the screen), click the mouse on the maximise icon appearing on the top right-hand corner of the worksheet (not the icon on the top right-hand corner of the Window). Click on icon to restore the original screen. If you click on the minimise icon, the workbook will be temporarily closed to its icon form (which will appear on the bottom left-hand corner of the screen). To open the worksheet again, click on the icon either or. If you click on the icon, the worksheet will be permanently closed. [**<u>You should always make sure that your work is saved before you close your document</u>**.]

1.3 The EXCEL Window

A number of things you see on the EXCEL Window need to be remembered throughout the EXCEL applications. These include the ***Title bar***, ***Menubar***, ***Standard toolbar***, ***Formatting toolbar***, ***Formula bar***, ***Cell reference*** (also known as ***Name box***), ***Active cell,***

Vertical scroll bar, Horizontal scroll bar, *the* ***Scroll arrows, Workbook name*** and ***Worksheet name*** (see Figure 1.1).

Figure 1.1 EXCEL Workbook Window

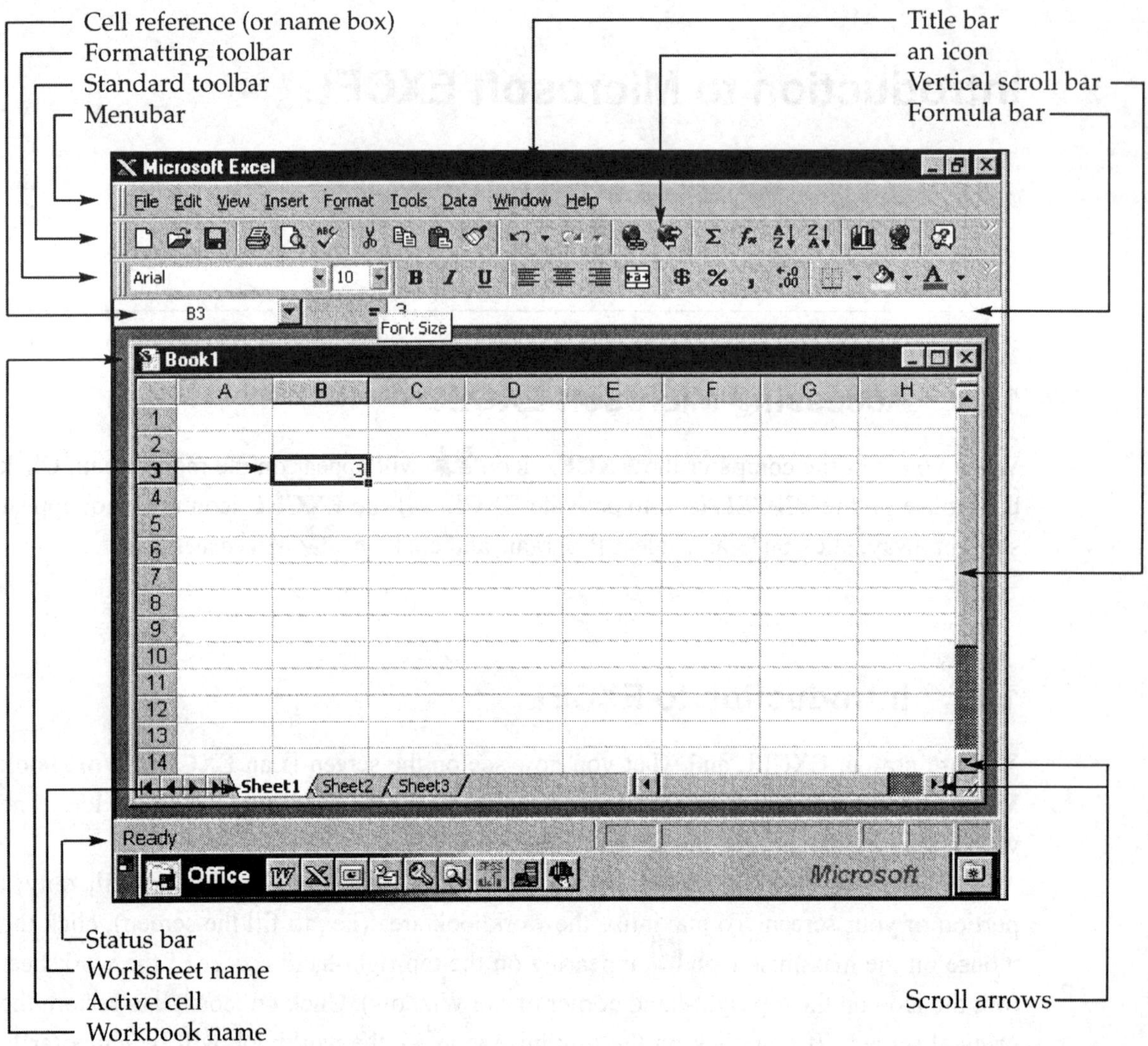

The Menubar

The *Menubar* is made up of a number of menus such as ***File, Edit, View, Insert, Format, Tools, Data, Window*** and ***Help.***. To select a menu on the *Menubar*, click the mouse on that menu. Each of these menus has a drop-down menu, which lists a number of commands. For example, Figure 1.2 shows what appears on the screen when we click on the ***File*** menu.

Figure 1.2 Menubar and the File menu

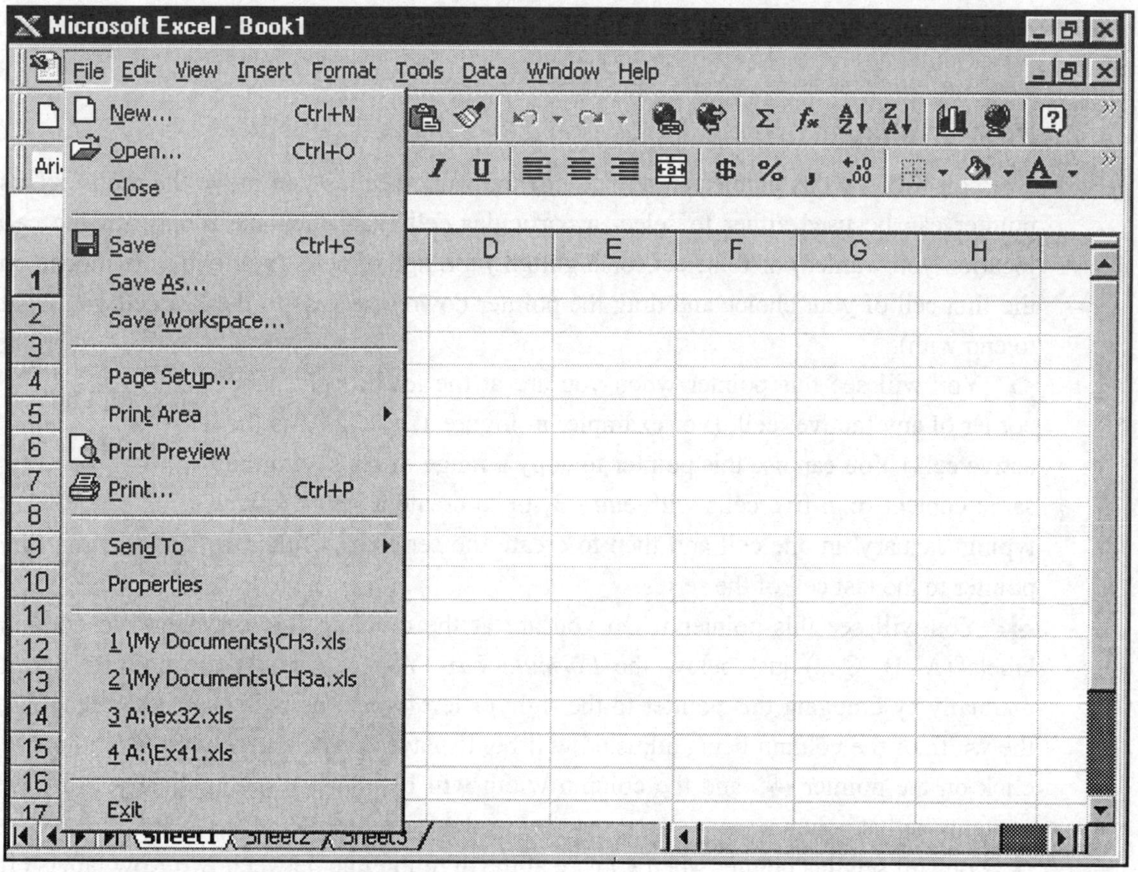

The Toolbars

EXCEL has a number of toolbars. You can see the list of all the available toolbars when you click on ***View*** on the *Menubar* and then on the *View* menu select *Toolbars...*. When you open an EXCEL worksheet, it normally displays two of the toolbars (these are highlighted on the menu with a ☑ against them), namely, the ***Standard toolbar*** and the ***Formatting toolbar*** below the *Menubar* (see Figure 1.1). Each toolbar contains a row of small rectangular and square buttons (which we call 'icons'). Some of these icons are shortcuts to the commands in the *Menubar*. For example, to open a file, instead of clicking the mouse on *File* on the *Menubar* and then selecting the *Open* command on the *File* menu (see Figure 1.2), we could straight away click the mouse on the icon **Open** (second square) on the ***Standard toolbar*** (see Figure 1.1).

You will also notice that when you move the mouse from one icon to another icon on the *Standard toolbar* or *Formatting toolbar*, EXCEL will show its operational name explicitly just below that icon. For example, when the mouse points to the icon Σ on the *Standard toolbar*, EXCEL will show its operation name as **AutoSum;** or the icon **B** on the *Formatting toolbar*, the operation name will be shown as **Bold**, and so on. In Figure 1.1, the mouse points to the font size **10** and the explicit name **Font Size** is displayed.

Before proceeding any further, we should be able to recognise some of the following common mouse pointer shapes in EXCEL.

Pointer Shapes

✜ You will see this pointer on any cell of the worksheet as you move the mouse. This pointer can be used either to select a particular cell (just click the mouse on the cell location you wanted) or to select (or highlight) a range of cells (you click the mouse on the first cell of your choice and drag the pointer down or across to the last cell you want to end with).

+ You will see this pointer when you are at the lower right hand corner of any 'active' cell. (For example, in Figure 1.1, cell B3 is the active cell.) You can use this pointer to copy a range of cells with the same content (e.g. five cells with entry 3) or to create a series (after typing 'January' in one cell and then to create the series of 12 months) by dragging the pointer to the last cell of the series.

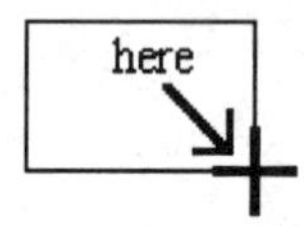

↔ You will see this pointer when you are at the dividing line between two column labels (A, B, C ...) just below the *Formula bar*. You can adjust the column width manually by dragging this pointer to the right or left (when the pointer is being dragged, the width of the column being adjusted will be displayed). Alternatively, you can double click on the pointer ↔ and the column width will be adjusted automatically to fit the contents in that column.

↕ You will see this pointer when you are at the dividing line between two Row labels (1, 2, 3 ...). You can adjust the row width by dragging this pointer upwards or downwards.

Some other useful keys and operations ......

1. To move from one cell to the other, you can either use the mouse to click the pointer ✜ or use the arrow keys →↑←↓ on the keyboard. If you want to move downwards only in the same column, you can also use the **Return** key.

2. To delete an entry in a cell, simply click the mouse on the cell and press the **Delete** button on the keyboard.

3. To highlight a number of cells (say, B1 to C2), click the mouse at the start cell B1 and drag the pointer ✜ to the end cell C2 (see Figure 1.3, where cells B1 to C2 are highlighted).

4. To delete an entire row (or an entire column), click any cell in that row (or column), click the mouse on **Edit** on the *Menubar* and then select the **Delete** option. Then in the *Delete dialogue box*, click the circle for **Entire Row** (or **Entire column**) and click **OK**. This operation will delete the entire row (or column) and move all the rows up by one row (or columns to the left by one column).

Figure 1.3 Highlighted Cells in a Worksheet

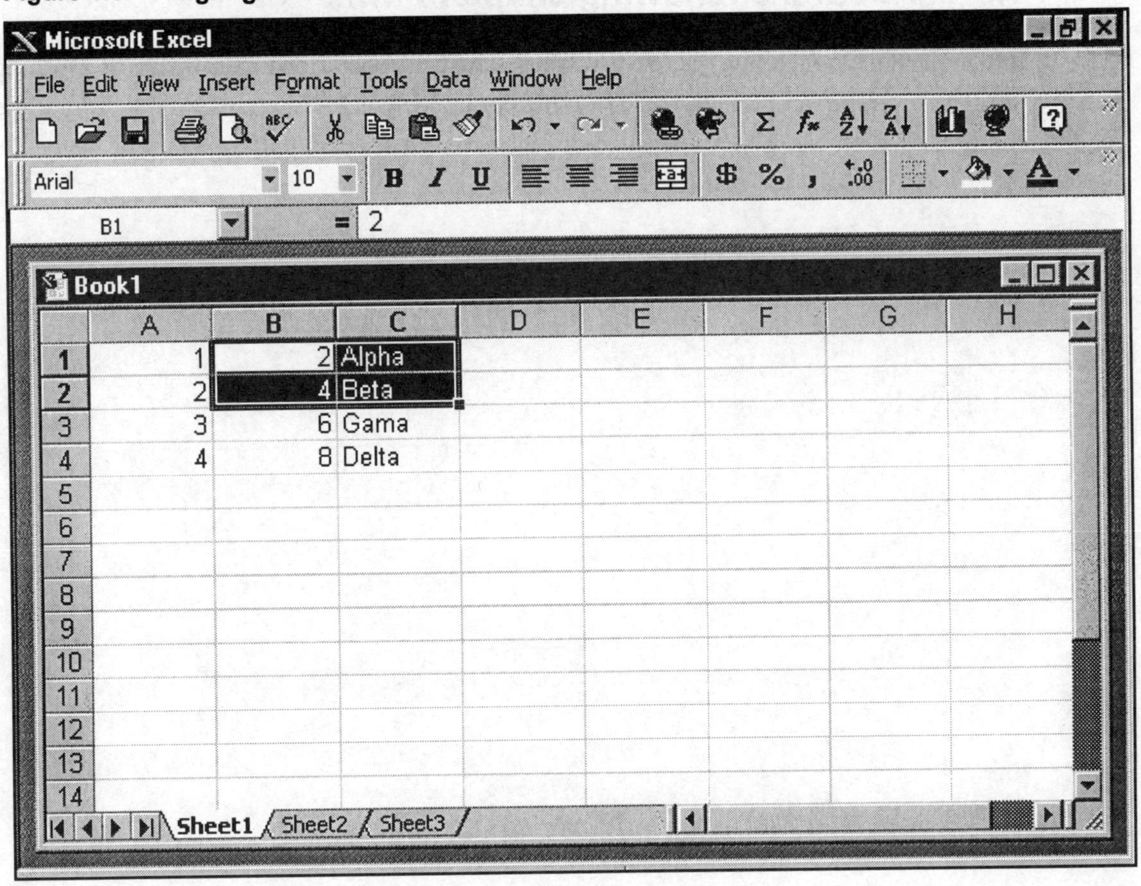

1.4 Online HELP facilities in EXCEL

EXCEL has a number of online help facilities at various levels which can be used to obtain information about most of the EXCEL commands/functions. There are several ways to access the online help in EXCEL. The three most commonly used methods are as follows:

1. Click the ? button (a question mark appearing at the end of the *Standard toolbar*). Type the topic name for which help is required, click **Search** and follow the instructions.
2. If help is required when you are in a dialogue box, click the ? button (appearing near the top right-hand corner) and drag it onto the particular area of the dialogue box for which help is required.
3. On the *Menubar*, select **Help** and, choose **Microsoft Excel Help** or press the **F1** key. Then type the topic name for which help is required.

1.5 About the following chapters ....

In the chapters that follow we learn various statistical operations in EXCEL by working through several examples. If you are not familiar with EXCEL, you should follow the solutions to each task given at the end of each example to learn these operations.

Chapter 2

Data Entry and Data Manipulation

In this chapter, we learn a number of file handling operations, file editing operations, operations for basic calculations, operations on sorting data sets and copying the text and tables of an EXCEL document into other Microsoft applications such as Microsoft Word.

Aim: At the completion of the examples in this chapter, you should be able to OPEN and SAVE a new spreadsheet, EDIT an existing spreadsheet, INSERT and DELETE rows and columns, COPY and MOVE data, use the SUM function and write simple FORMULAS to CALCULATE new variables. You should also be able to carry out some simple spreadsheet formatting including changing the width of columns; centering headings; bolding, italicising and underlining text; and changing the font size of the text, sorting data, and copying EXCEL documents into Microsoft Word documents.

2.1 Example 2.1: Entering data, formatting and performing simple calculations

Table 2.1 below contains information on petrol sales of the BPP Service Station at Sunnybank, Queensland, for each month in 1997. The manager wishes to use this information to get an idea of the total cost, total revenue and total net income for the year. Use EXCEL to perform the following tasks.

Task 1 (i) Enter the data as it appears in the table onto a new worksheet but with the normal font size (10) starting from cell A1.

(ii) As you would have noticed all table headings in the first four lines are bold faced; Heading 1 is of font size 14; Headings 2 and 3 are of font size 12; and all others are of font size 10. In addition to the above, the first two

TABLE 2.1

SERVICE STATION BPP

MONTHLY PETROL SALES 1997

****** (Delete me)**

Month	**Sales (thousands of litres)**	**Average Price at the Pump ($)**	**Average Supply Price ($)**	**Net Income ($)**
January	4000	.55	.40	
February	4500	.57	.41	
March	4000	.56	.39	
April	3600	.57	.40	
May	4100	.55	.41	
June	3800	.58	.40	
July	3900	.60	.40	
September	4000	.62	.41	
October	4100	.63	.39	
November	4200	.62	.38	
December	4300	.60	.40	

headings are underlined, and the last heading is in italics. Make the necessary changes to the headings in your worksheet to look like the above.

(iii) As you would have noticed, some columns are not wide enough to include all the information, while others may be narrowed. Adjust the column widths as needed.

(iv) The table headings could be centred. Centre the headings TABLE 2.1, SERVICE STATION BPP and *MONTHLY PETROL SALES 1997* across all used columns.

(v) When you type the entry **.60**, you would have noticed that it appeared as .6 whereas .55 would appear as .55. Make all numbers in columns C and D to 2 decimals.

Task 2 Save the worksheet onto your disk (in drive A) using the filename A:EX21.XLS and close the file.

Task 3 Now open the file A:EX21.XLS by selecting the open command on the **File** menu on the *Menubar*.

(i) As you would have noticed, figures for August have been left out. Insert a new row between July and September and enter the following figures for August.

Sales (thousands of litres) = 3950

Average Price at the Pump = $.61

Average Supply Price = $.39

(ii) Delete the entire 4th row starting with ****.

Task 4 Calculate the total petrol sales (thousands of litres) for 1997 in two ways using EXCEL (either by using the icon [Σ] **Autosum** or by typing the formula).

Task 5 The manager wants to know the total revenue for the year 1997.

Revenue = Sales × Average Price at the pump

(i) Insert a column between the pump price and supply price columns and give it a heading 'Revenue'.

(ii) In the January cell under the heading 'Revenue', type the above formula to calculate the revenue and copy this to cover all months. Calculate the 'Total Revenue' for the year 1997

Task 6 The manager also wants to calculate his total cost for the petrol sold.

Cost = Sales × Average Supply Price

(i) Move the column heading 'Net Income ($)' across one column to the right, and replace it with the heading 'Cost ($)'.

(ii) In the January cell under the heading 'Cost', type the formula to calculate the 'Supply Cost' using petrol sales and supply price and copy this to cover all months. Calculate the 'Total Cost' for the year 1997.

Task 7 Calculate the Net Income for each month and for the year 1997.

Net Income = Revenue - Cost

Task 8 Set $ sign to all the dollar values in columns C to G. Centre the column headings.

Task 9 Save the worksheet onto your disk with the same filename.

Task 10 Print a hardcopy of your final spreadsheet.

Solution to Example 2.1

When you start EXCEL, a worksheet of a new workbook is already present. Each data point should be entered into a separate cell. Every cell on the worksheet has an address identified by its column (alphabet) and row (number) locations. For example, the third cell in the H column will be identified as H3. The active cell in use will be highlighted with dark black boundaries (see Figure 1.1) and its location can be seen in the cell reference area. When you move from one cell to the other (click the mouse on a new cell) you will notice the location reference will change in the cell reference area automatically.

Task 1 (i) Type the data exactly as it appears in Table 2.1 on the worksheet.

Note: To enter the months in column A, after typing 'January' in cell A8, you can create the series for the twelve months automatically by moving the ✚ pointer to the lower righthand corner of the cell and drag the + pointer down 12 rows, instead of typing all months manually. [In this task, as you do not have August, if you generate months automatically then you have to delete the entire row for August. See below Task 3(ii) on how to delete a row].

(ii) The heading 'TABLE 2.1' is currently in font size 10. To increase its font size to 14, first you highlight that heading and click the mouse on the font size icon arrow ▼ on the *Formatting toolbar* and then select the font size 14 (see Figure 2.1). Do the same for the other headings to change them to font size 12. To bold the headings, highlight the headings and then click on the icon **B** **Bold** of the *Formatting toolbar*. To underline a heading, first highlight it and then click on the icon **U** **Underline** on the *Formatting toolbar*. Similarly, use *I* for **Italics** from the *Formatting toolbar*.

(iii) When you type the headings you would have noticed that it will spill over columns B and C as the column A width is not large enough. To increase the width of column A, move the ✥ pointer to the dividing line between column labels A and B. At the dividing line, you will see the pointer look like ↔. Now drag this pointer to the right to increase the column width. Do the same with the other headings in order to see them in full. You can do the same using the pointer ↕ if you want to increase the row height. Alternatively, you can double click the pointer ↔ and the column width will be adjusted automatically to the widest entry in the column.

(iv) To centre the table heading 'TABLE 2.1'across all used columns, highlight the cells by clicking the mouse on cell A1 and dragging it up to cell E1. Release the mouse and select the icon **Centre Across Columns** from the *Formatting toolbar*. Do the same for the next two headings. Now re-adjust the width of column A as in (iii) above.

(v) When you type the entries for the price series in column C, you will notice that 0.60 will appear only as 0.6 whereas .55 will appear as 0.55. To have all numbers in two decimals (for example, 0.55, 0.60, 0.56, and so on), highlight these cells and click on the icon **Increase decimal** from the *Formatting toolbar*.

Now your worksheet should look like Figure 2.1.

Task 2 After typing all the data, to save the file, you have two methods:

Method 1 Click the mouse on **File** on the *Menubar* and select option **Save** or **Save as** (see Figure 1.2).

Method 2 Click on the icon **Save** (third from the left) on the *Standard toolbar* (see Figure 1.1).

Under both methods a *Save As dialogue box* will appear on the screen. Insert your disk in Drive A, click on the **File name:** rectangle, type **A:filename.XLS** (for example, **A:EX21.XLS**), and then click on **Save**. When you have completed this sequence, you will be back in the worksheet. Now click on **File** on the *Menubar* and select option **Close** to close the file.

Figure 2.1 Formatting the EXCEL worksheet

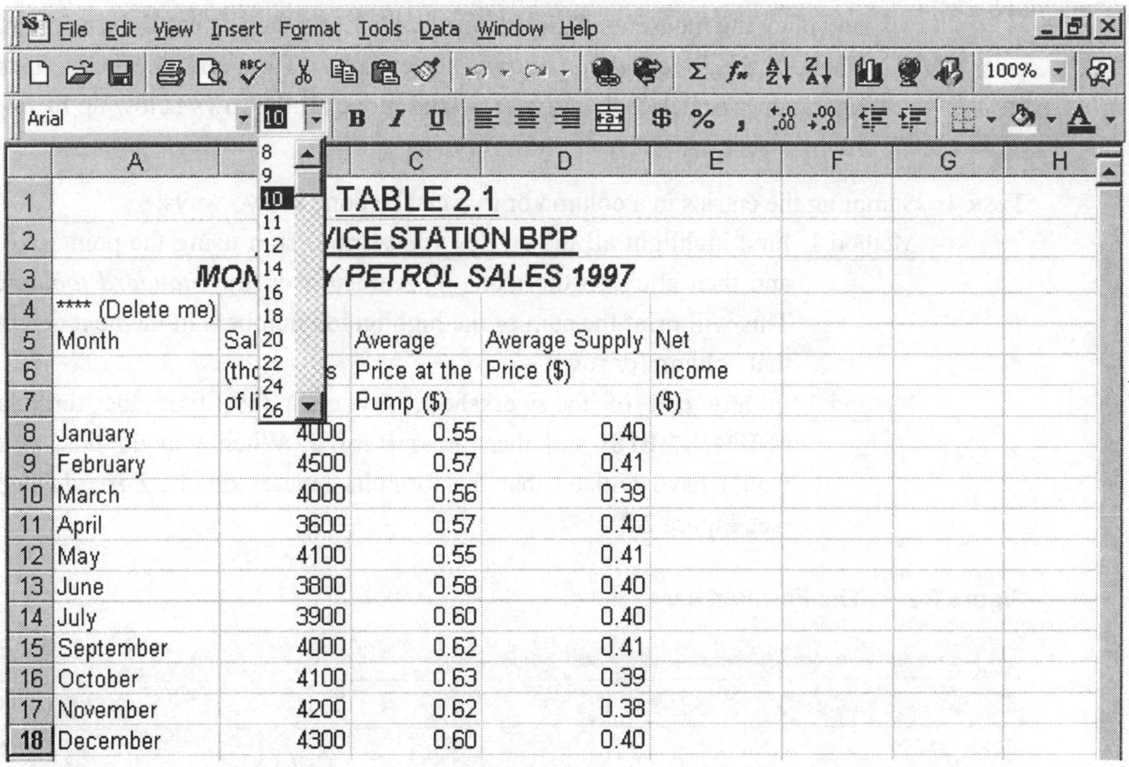

	A	B	C	D	E	F	G	H
1			TABLE 2.1					
2			VICE STATION BPP					
3	MON		Y PETROL SALES 1997					
4	**** (Delete me)							
5	Month	Sal	Average	Average Supply	Net			
6		(thc	Price at the	Price ($)	Income			
7		of li	Pump ($)		($)			
8	January	4000	0.55	0.40				
9	February	4500	0.57	0.41				
10	March	4000	0.56	0.39				
11	April	3600	0.57	0.40				
12	May	4100	0.55	0.41				
13	June	3800	0.58	0.40				
14	July	3900	0.60	0.40				
15	September	4000	0.62	0.41				
16	October	4100	0.63	0.39				
17	November	4200	0.62	0.38				
18	December	4300	0.60	0.40				

Task 3 (i) Now to insert the August figures in the worksheet, we have to edit the file A:EX21.XLS'. To do this, we first open the file A:EX21.XLS. There are two methods to do this:

Method 1 Click the mouse on **File** on the *Menubar* and then select option **Open**.

Method 2 Click the mouse on the icon **Open** (second from the left) on the *Standard toolbar*.

Under both methods an *Open dialogue box* will appear on the screen. There are two methods to specify the filename:

Method 1 Click on the arrow in the *Look in:* box and click **3½ Floppy (A:)**. Then click **EX21.XLS** from the filename list box, and then click **Open**.

Method 2 Click the mouse on the *File name:* rectangle and type **A:EX21.XLS** and click **Open**.

Now to insert a row for August, you first select any cell in the September row and then click the mouse on **Insert** on the *Menubar* and select the **Rows** option. This will make a blank row between the July and September rows. Now you type the information for August.

(ii) To delete the fourth heading, '**** delete me', click any cell in that row and click the mouse on **Edit** on the *Menubar* and then select **Delete** option. Then, in the *Delete dialogue box*, select **Entire Row** and click **OK**. This operation will delete the entire row and move all the rows below up by one row.

Task 4 Summing the entries in a column or row can be done in two ways:

Method 1 First highlight all the entries you want to sum using the pointer ✥ and then click on the icon Σ **Autosum** on the *Standard toolbar*. This will print the sum of the highlighted numbers in the next cell of that column (or row).

Method 2 In any cell of the worksheet, say cell B19, type the formula **=SUM(B7:B18)** and then press **Return**. When you do this, you would have noticed that the formula appears on the *Formula bar* (see Figure 2.2).

Figure 2.2 The Formula bar

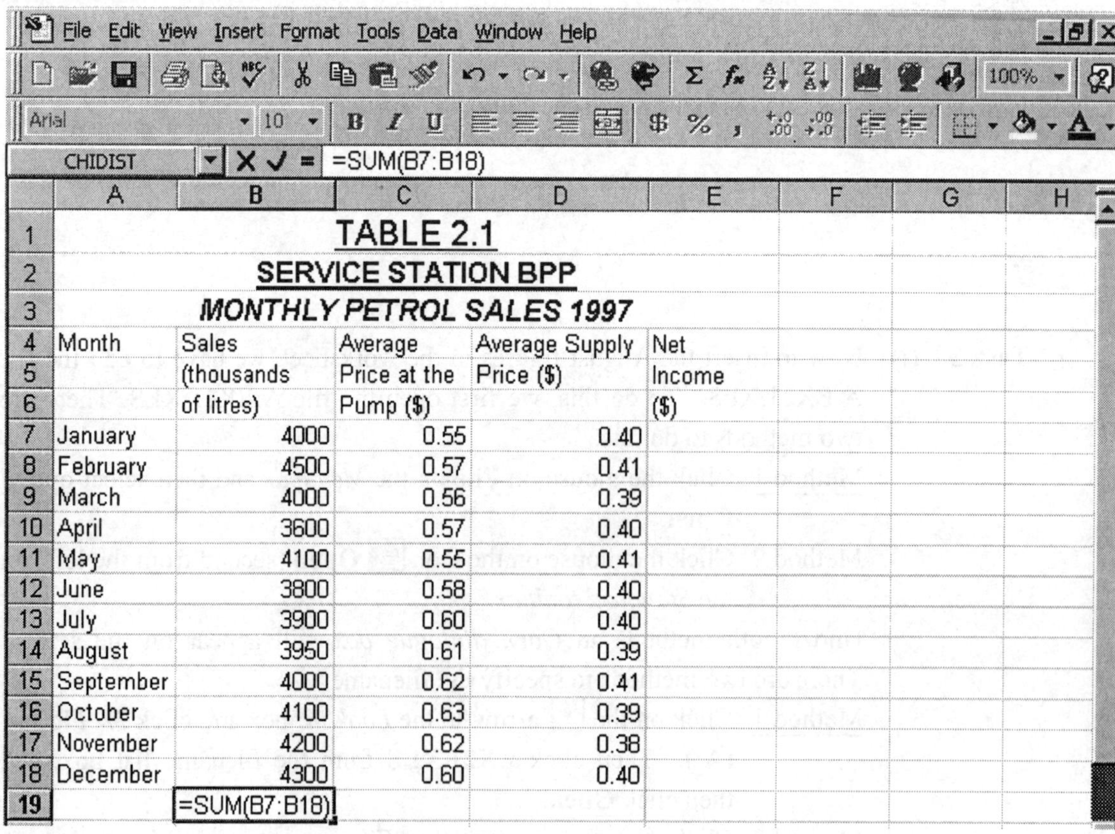

	A	B	C	D	E
4	Month	Sales	Average	Average Supply	Net
5		(thousands	Price at the	Price ($)	Income
6		of litres)	Pump ($)		($)
7	January	4000	0.55	0.40	
8	February	4500	0.57	0.41	
9	March	4000	0.56	0.39	
10	April	3600	0.57	0.40	
11	May	4100	0.55	0.41	
12	June	3800	0.58	0.40	
13	July	3900	0.60	0.40	
14	August	3950	0.61	0.39	
15	September	4000	0.62	0.41	
16	October	4100	0.63	0.39	
17	November	4200	0.62	0.38	
18	December	4300	0.60	0.40	
19		=SUM(B7:B18)			

Note: Suppose we want to know the total for the last 10 months, the formula should be =SUM(B9:B18). To make this change, edit the formula by clicking on the formula on the *Formula bar* at 7 and delete 7 and type **9** and then click on the green colour icon ☑ on the *Formula bar* to save the correction to the formula. Now the revised formula will appear as =SUM(B9:B18). You can also edit the formula on the cell by going to that particular cell, press function key **F2** , delete 7 and type **9**. Press the **Enter** key when done with the editing.

Task 5 (i) To insert a column between the 'Average Price at the Pump ($)' in column C and the 'Average Supply Price ($)' in column D, click on any cell of column D and then click the mouse on **Insert** on the *Menubar* and then select **Columns** option. This will make a new column D and the old column D will become column E and old column E will become column F. Type the heading, 'Revenue ($)' in cell D4 for this new column D.

(ii) To calculate the revenue, we use the formula:

$$\text{Revenue} = \text{Sales} \times \text{Average price at pump}$$

To calculate the revenue for January, in cell D7 under the heading 'Revenue ($)', type **=B7*C7** and then press Enter.

(iii) To fill the other cells in this column, simply drag the pointer + from the lower right-hand corner of cell D7 down to the December cell, D18. To calculate the total revenue, simply highlight all cells D7 to D18 and use the icon Σ **Autosum** on the *Standard toolbar* or type the formula **=SUM(D7:D18)** in cell D19.

Task 6 To move the 'Net Income ($)' heading across one column to the right, insert a column between columns E and F as discussed in Task 5(i). In cell F4, type the new heading 'Cost ($)'. To calculate the monthly and total costs, use the formula

$$\text{Cost} = \text{Sales} \times \text{Average Supply Price}$$

and follow instructions similar to Task 5.

Task 7 In the January cell under the heading 'Net Income ($)' (i.e. in cell G7), type the formula **=D7-F7** and press **Return**. Then to fill the other cells in column F, simply drag the pointer + from the lower right-hand corner of cell G7 down to cell G18. To obtain the total Net Income for the year 1997, highlight cells G7 to G18 and then click on icon Σ **Autosum** from the *Standard toolbar*.

Task 8 To put $ sign to all the entries in cells C7 to G19 in columns C to G, first highlight all entries in these columns and then click on icon $ **Currency Style** on the *Formatting toolbar*. If you see anything like this ###### in any cell, it means that the 'cell width' is not enough to accommodate the contents. To fix this, increase the column width as discussed earlier. To centre the column headings in cells A5 to E7, highlight cells A5 to E7 and click on the icon **Centre** on the *Formatting toolbar*.

When you have completed all the above tasks, the worksheet should look like Figure 2.3.

Figure 2.3 Final EXCEL output

	A	B	C	D	E	F	G
1				TABLE 2.1			
2				SERVICE STATION BPP			
3				*MONTHLY PETROL SALES 1997*			
4	Month	Sales	Average	Revenue ($)	Average Supply	Cost ($)	Net
5		(thousands	Price at the		Price ($)		Income
6		of litres)	Pump ($)				($)
7	January	4000	$ 0.55	$ 2,200.00	$ 0.40	$ 1,600.00	$ 600.00
8	February	4500	$ 0.57	$ 2,565.00	$ 0.41	$ 1,845.00	$ 720.00
9	March	4000	$ 0.56	$ 2,240.00	$ 0.39	$ 1,560.00	$ 680.00
10	April	3600	$ 0.57	$ 2,052.00	$ 0.40	$ 1,440.00	$ 612.00
11	May	4100	$ 0.55	$ 2,255.00	$ 0.41	$ 1,681.00	$ 574.00
12	June	3800	$ 0.58	$ 2,204.00	$ 0.40	$ 1,520.00	$ 684.00
13	July	3900	$ 0.60	$ 2,340.00	$ 0.40	$ 1,560.00	$ 780.00
14	August	3950	$ 0.61	$ 2,409.50	$ 0.39	$ 1,540.50	$ 869.00
15	September	4000	$ 0.62	$ 2,480.00	$ 0.41	$ 1,640.00	$ 840.00
16	October	4100	$ 0.63	$ 2,583.00	$ 0.39	$ 1,599.00	$ 984.00
17	November	4200	$ 0.62	$ 2,604.00	$ 0.38	$ 1,596.00	$ 1,008.00
18	December	4300	$ 0.60	$ 2,580.00	$ 0.40	$ 1,720.00	$ 860.00
19		48450		$ 28,512.50		$ 19,301.50	$ 9,211.00

Task 9 When you have completed all these tasks, to save the file with the same filename, use one of the following two methods:

Method 1 Click on **File** on the *Menubar* and select command **Save**.

Method 2 Click on the icon **Save** (third from the left) on the *Standard toolbar*.

Task 10 To print the final workbook with all the tasks completed, either click on the icon **Print** (fourth from the left) on the *Standard toolbar* or click the **File** menu on the *Menubar*. In most cases where the file can be fitted into a portrait A4 width, select **Print...** (see Figure 1.2) and on the Print menu, click **OK**. In this example, as the width is too large, we need to print it in the landscape mode. To change to landscape mode, first select **Page Setup...** from the **File** menu. Click on the circle for **Landscape** (see Figure 2.4) and then click **Print**... and on the Print menu, click **OK**.

Figure 2.4 Page Setup dialogue box

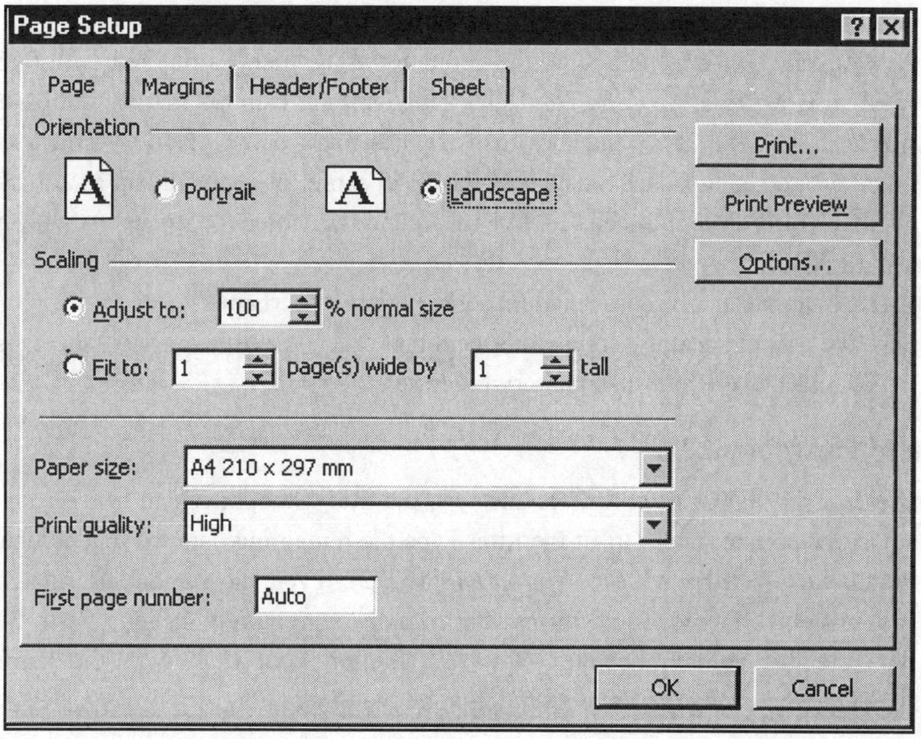

2.2 Example 2.2: Performing simple calculations and sorting data

The table below shows the marks achieved by a sample of 10 students in a statistics subject where the total number of enrolments is 100.

Student ID	Surname	Computer Assignment (out of 25)	Mid-Semester Exam (out of 40)	Final Exam (out of 100)	Total Score (out of 100%)
6510	Griffith	24	35	80	
4616	Chessman	16	24	72	
6724	Cattleman	21	40	90	
9245	Patrick	7	5	37	
1367	Yeo	18	21	40	
2493	Mason	22	16	48	
4765	Marshall	19	30	74	
5234	Gardner	5	14	32	
6556	Keating	15	29	71	
7272	Howard	20	18	69	

Use EXCEL to perform the following tasks.

Task 1 Enter the data as above in cell A1 to E13. In cells F1 to F3 type the last column heading. Save the data in a file with the name 'A:EX22.XLS'.

Task 2 As you can see from the above, the computer assignment is marked out of 25 *(while it is worth 15% of the Total)*, the mid-semester exam is marked out of 40 *(while it is worth 25%)* and the final exam is marked out of 100 *(while it is worth 60%)*. Using EXCEL calculate the Total score for each student as a mark out of 100 and store it in cells F4 to F13. Round the scores to the nearest integer using EXCEL.

Task 3 Sort the sample exam scores in alphabetical order by surname.

Task 4 Sort the sample exam scores in ascending order. Save your worksheet as 'A:EX22.XLS' on your disk and print the file.

Solution to Example 2.2

Task 1 Create a new EXCEL worksheet and type the title 'Student ID' in cell A1 and adjust the column width to fit the title. Type the remaining data for that column in cells A4 (6510), A5 (4616) ---, A13 (7272). Type the remaining data in columns B, C, D and E and format the columns as required. In cells F1 to F3 type the last column heading. Save the file on your disk with the name **A:EX22.XLS** using the **Save As...** option on the **File** menu.

Task 2 To calculate the Total score we have to write our own mathematical formula. We need to convert

(1) computer assignment scores to *out of 15%* by multiplying entries in cells C4 to C13 by 15/25.

(2) mid-semester scores to *out of 25%* by multiplying entries in cells D4 to D13 by 25/40.

(3) final exam score to *out of 60%* by multiplying entries in cells E4 to E13 by 60/100.

and then add the three scores to obtain the Total score.

For example, in cell F4, type the formula for the Total score, **=C4*(15/25)+D4*(25/40)+E4*(60/100)** and then copy this formula onto cells F5 to F13. Highlight the column and round off to the nearest integer using the icon **Decrease Decimal** on the *Formatting toolbar*. The results are given in Figure 2.5.

Task 3 First highlight the data to be sorted, cells A4 to F13. To sort the data by a column, click on **Data** on the *Menubar* (see Figure 2.6) and select **Sort....** A *Sort dialogue box* will appear on the screen. In this dialogue box, under **Sort by** option, click on the arrow and since we want to sort the data by surname, select **Column B** and then click **OK** (see Figure 2.7). The resulting table will look like Figure 2.8.

Figure 2.5 Total scores

	A	B	C	D	E	F
1	Student ID	Surname	Computer	Mid-Semester	Final	Total
2			Assignment	Exam	Exam	Score
3			(out of 25)	(out of 40)	(out of 100)	(out of 100%)
4	6510	Griffith	24	35	80	84
5	4616	Chessman	16	24	72	68
6	6724	Cattleman	21	40	90	92
7	9245	Patric	7	5	37	30
8	1367	Yeo	18	21	40	48
9	2493	Mason	22	16	48	52
10	4765	Marshall	19	30	74	75
11	5234	Gardner	5	14	32	31
12	6556	Keating	15	29	71	70
13	7272	Howard	20	18	69	65

Figure 2.6 Menubar and the Data menu

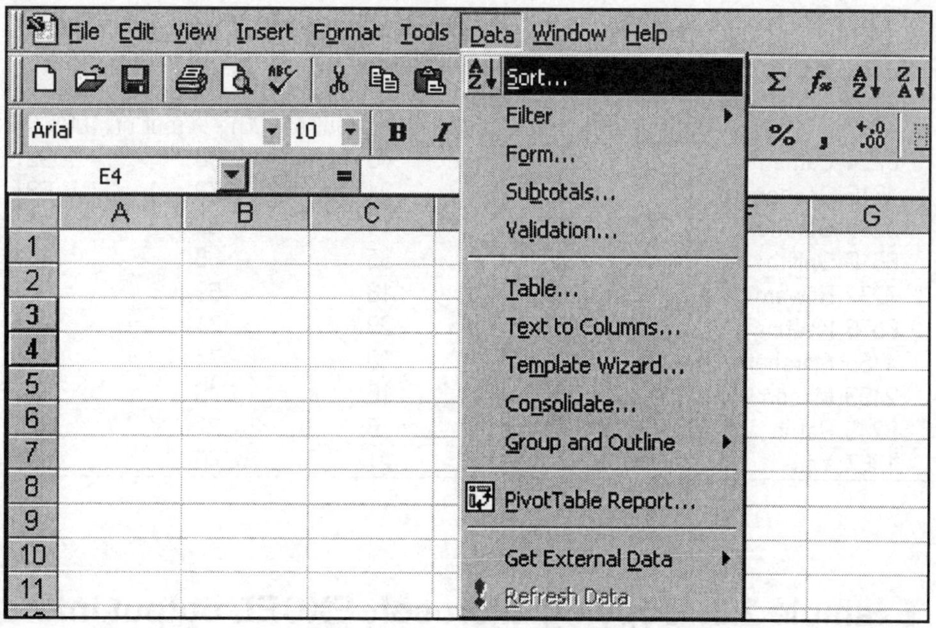

Task 4 To sort the data according to the exam scores in ascending order, repeat Task 3 with the selection of **Column F**. Save the EXCEL workbook with the same filename **A:EX22.XLS** as in Task 1, using the **Save** option on the **File** menu.

Figure 2.7 SORT dialogue box

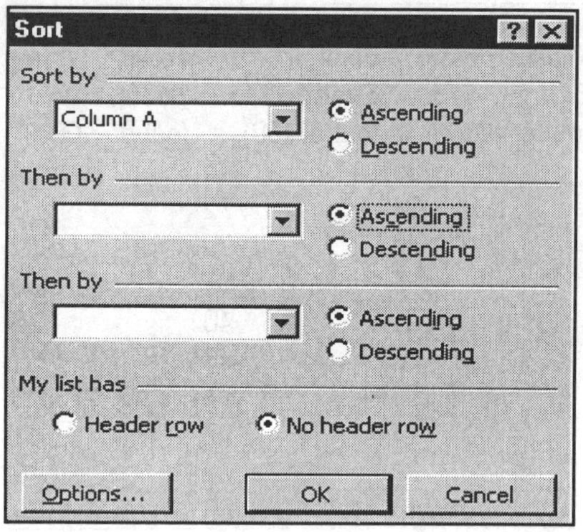

Figure 2.8 Total scores in alphabetical order

	A	B	C	D	E	F
1	Student ID	Surname	Computer	Mid-Semester	Final	Total
2			Assignment	Exam	Exam	Score
3			(out of 25)	(out of 40)	(out of 100)	(out of 100%)
4	6724	Cattleman	21	40	90	92
5	4616	Chessman	16	24	72	68
6	5234	Gardner	5	14	32	31
7	6510	Griffith	24	35	80	84
8	7272	Howard	20	18	69	65
9	6556	Keating	15	29	71	70
10	4765	Marshall	19	30	74	75
11	2493	Mason	22	16	48	52
12	9245	Patric	7	5	37	30
13	1367	Yeo	18	21	40	48

2.3 Example 2.3: Copying Microsoft EXCEL output into a Microsoft Word document

Use the data from Example 2.2 and EXCEL to perform the following tasks.

Task 1 Copy the output in Figure 2.5 from Example 2.2 (saved in EXCEL file 'A:EX22.XLS') into a new Word document.

Task 2 Save the document on your disk as 'A:EX22.DOC'.

Solution to Example 2.3

Task 1 One of the important features of Microsoft applications is that the output from one application (e.g., part of an EXCEL output) can be copied into another application (e.g., into a Word document). To copy the output in Figure 2.5, produced in Example 2.2 in EXCEL, into a word document, we perform the following steps:

Step 1 Open the EXCEL file **A:EX22.XLS**.

Step 2 Highlight the total area of the output in your worksheet using the mouse.

Step 3 Click on the icon **Copy** on the *Standard toolbar*, or click on the **Edit** menu on the *Menubar* and select **Copy**.

Step 4 Click the mouse on the icon **Microsoft Word.**

Step 5 Click the mouse on the position of the Word document where the table is to be copied.

Step 6 Click on the icon **Paste** on the *Standard toolbar* or click on the **Edit** menu on the *Menubar* and then select **Paste**. Now you will find the output in Figure 2.5 from the EXCEL document inserted into the Word document.

Task 2 To save the Word document, click on the icon **Save**, or click on the **File** menu on the *Menubar* and select **Save,** or **Save As...** and, type the filename **A:EX22.DOC**. *[Note that the extension for the Word document is '.DOC'. If you are not careful, you may overwrite the file 'EX22.XLS' with the new document and will loose all the output from Example 2.2].*

2.4 Exercises to Chapter 2

Exercise 2.1

The following table shows the annual expenditure patterns of Australians on beer, wine and spirits and the total annual expenditure on all consumer goods for the period 1985 to 1992. Use EXCEL to perform the following tasks.

Task 1 (i) Enter the above data as it appears in 'Table EX2.1' onto a new worksheet but with the normal font size (10) starting from cell A1.

(ii) As you would have noticed all headings in the first three lines are bold faced; Heading 1 is of font size 14; Headings 2 and 3 are of font size 12; and all others are of font size 10. Also the third heading is in italics and underlined. Make the necessary changes to the headings in your worksheet to look like the above.

(iii) As you would have noticed in your worksheet, some columns are not wide enough to include all the information, while others may be narrowed. Adjust the column widths as needed.

(iv) Centre the headings across all used columns.

Task 2 Save the worksheet onto your disk (in drive A) using the name A:ALC.XLS and close the file.

TABLE EX2.1

Expenditure on Beer, Wine, Spirits and All Goods, Australia, 1985-1992 (Current dollars per capita)

Year	Beer	Wine	Spirits	Total Expenditure on All Goods
1985	249.33	101.54	68.94	8058.17
1986	273.78	104.31	80.72	8863.55
1987	291.03	111.51	83.54	9697.11
1988	316.09	116.13	94.14	10632.20
1989	328.10	116.13	105.57	11588.10
1991	360.96	113.09	119.80	13219.29
1992	351.84	122.19	118.54	13684.98
1996				
Mean				

Task 3 Now open the file A:ALC.XLS.

(i) As you would have noticed, figures for 1990 have been left out. Insert a new row and enter the following per capita expenditure figures (in dollars) for 1990.

Beer = 344.12
Wine = 115.94
Spirits = 115.55
All goods = 12701.13

(ii) Delete the row for the year 1996.

Task 4 Insert a new column with heading 'Total Alcohol' between the 'Spirits' and 'Total Expenditure on All Goods' columns. Write the appropriate formula to calculate the per capita annual expenditure on alcohol using EXCEL for the years 1985 to 1992 and enter them in the new column.

Task 5 A social worker is interested in knowing 'on average, how much money did an Australian spend annually on beer, wine and spirits, total alcohol and on all goods during the period 1985 to 1992?'. Use the function icon 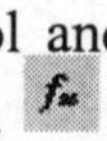 on the *Standard toolbar* and the function AVERAGE to calculate the entries in the row marked 'Mean'.

Task 6 A marketing consultant wants to know 'what proportion of their total alcohol expenditure was allocated to beer, wine and spirits?' and 'what proportion of the total expenditure on all goods was allocated to beer, wine, spirits and to total alcohol?' during 1985 to 1992. Set up a new table with the following headings:

TABLE EX2.1A

Expenditure Shares: Beer, Wine, Spirits and Total Alcohol Australia, 1985-1992 (in percentages)

Now, setup the table with the following column headings:

	Shares within alcohol			Shares among all goods			
Year	Beer	Wine	Spirits	Beer	Wine	Spirits	Total alcohol

Then, type years 1985, ….., 1992 in the first column. Use the appropriate formula to fill the various columns of the table. When you have completed, for example, the entries for 1985 should be the following:

1985 59.39 24.19 16.42 3.09 1.26 .86 5.21

Task 7 Set dollar ($) signs to all the dollar values in Table EX2.1.

Task 8 Insert page breaks between the two tables.

Task 9 Save your worksheet on your disk and print the file.

Task 10 Copy Table EX2.1 and Table EX2.1A into a new word document and save the word document as 'A:ALC.DOC' on your disk. Print the Word document.

Exercise 2.2

The following are the average scores given to the short-listed contestants by a panel of judges in the final stages of the *Miss International 1996 Competition*. The total score for each contestant was calculated based on 3 sub-scores: Interview on Stage which is worth 20% of the Total score; Appearance in their National dress worth 50% and General appearance worth 30%.

TABLE EX2.2

Average Scores: Miss International 1996

Country of Origin	Interview on Stage (out of 10)	Appearance in National Dress (out of 20)	General Appearance (out of 100)	Total Score (out of 100%)
USA	8.7	15.5	85	
Kenya	6.9	18.2	77	
India	7.0	20.0	79	
Canada	7.6	16.3	78	
Germany	8.9	17.1	83	
France	8.6	16.9	82	
Australia	8.3	16.7	91	
UK	8.4	16.5	86	

If you were the Statistical consultant hired by the *Miss International 1996 competition* organisers, use EXCEL to perform the following tasks to calculate the Total Score out of 100% and select the *Miss International 1996.*

Task 1 Enter the data as above in cell A1 to D14. In cells E4 and E6, type the last column heading. Save the data in a file with filename 'A:MSINT.XLS'.

Task 2 As you can see from the above, the score for the interview on stage category is awarded out of 10 *(while it is worth 20% of the Total score)*, the appearance in National dress category is awarded out of 20 *(while it is worth 50%)* and the general appearance category is awarded out of 100 *(while it is worth 30%).* Calculate the Total score out of 100% for each short-listed contestant and store it in cells E7 to E14. Round the scores to the nearest integer using EXCEL.

Task 3 Sort the short-listed contestants by country in Alphabetical order.

Task 4 Sort the short-listed contestants by Total score in descending order. Who is Miss International 1996? Who is the runner-up in the competition?

Task 5 Save your output in file 'A:MSINT.XLS' and print the file.

Chapter 3

Graphical Descriptive Methods: Summarising Data in Tables and Graphs

Descriptive statistics involves methods of arranging, summarising, and presenting essential information contained in a set of data. There are a number of graphical and numerical descriptive methods that can be used for this purpose. This chapter focuses on the graphical methods of summarising and presenting data.

The most popular way of summarising a *quantitative data* set is to group the data to form a *frequency distribution* or a *relative frequency distribution*. The graphical presentation of the frequency (or relative frequency) distribution is called a *histogram* or a *frequency polygon*, which convey the shape of the distribution of the data graphically.

A *qualitative data* set can be displayed graphically using a *pie chart* or a *bar chart*. Frequency distributions of quantitative data can also be displayed graphically using pie charts or bar charts, however, this is not very common. The type of chart to be used in a particular situation depends on the particular information the user wants to emphasise.

3.1 Frequency distributions and histograms

Frequency distribution is a table that groups a raw data set into non-overlapping intervals called *classes* or *class intervals* and records the number of observations in each class, called the *frequency* or *absolute frequency* of that class. Frequencies are graphically represented by either a *histogram* or a *frequency polygon*. *Relative frequency* of a class is calculated by dividing the absolute frequency of the class by the total frequency of all classes; and *cumulative frequency* of a class is obtained by adding the frequency of that class to the frequencies of all previous classes. The graph of the cumulative relative frequencies is called an *Ogive* curve.

In EXCEL, there are two starting points to construct a histogram of a data set. The first situation is when we have access to the raw data. The second is when we have access only to the frequency (or relative frequency) distribution table and not to the raw data.

To construct a frequency distribution from raw data, we first select **Tools** on the *Menubar* (see Figure 3.1) and then select **Data Analysis...**[1] from the available options. This will result in the *Data Analysis dialogue box* (see Figure 3.2). From this dialogue box, we select the option, **Histogram**, which will give another dialogue box. Based on the options selected from the *Histogram dialogue box* (see Figure 3.3), for a given data set, EXCEL will construct a frequency histogram, a relative frequency histogram, a cumulative relative frequency histogram or an Ogive curve.

Figure 3.1 Menubar and the Tools menu

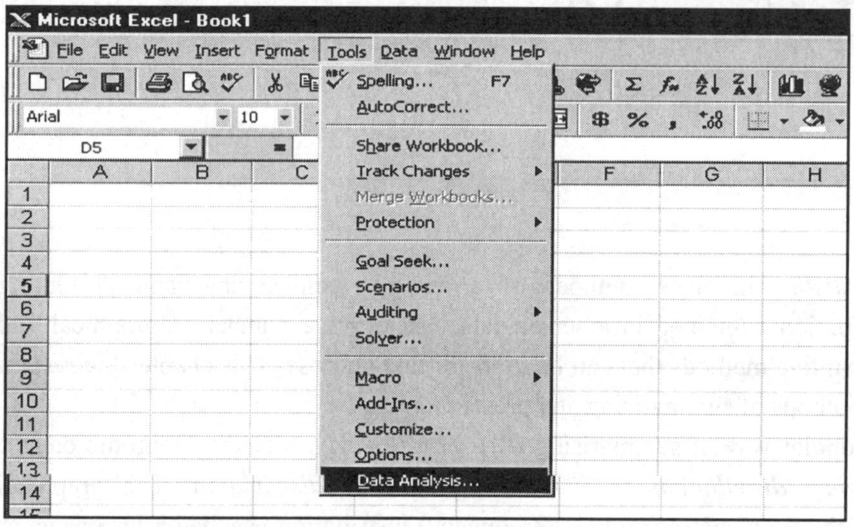

Figure 3.2 Data Analysis dialogue box

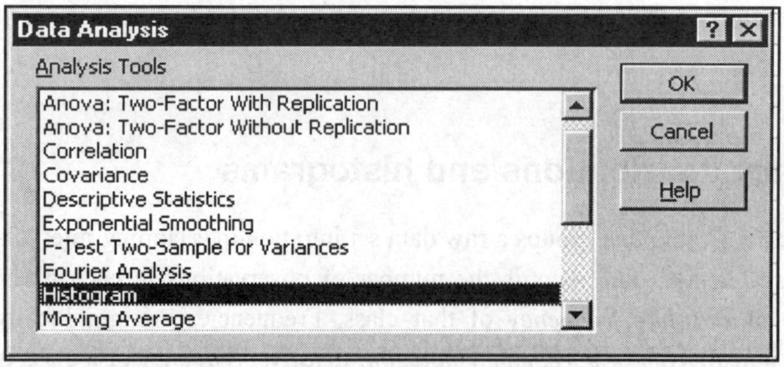

1 If you cannot find **Data Analysis...** on the *Tools* menu, to add it to the *Tools* menu first click on **Add-Ins...** on the *Tools* menu which will display another menu. Click the mouse on the little square against **Analysis Toolpak** (a tick, ☑, will appear inside the square) and then click on **OK**. Now, if you go back to the *Tools* menu, you will find **Data Analysis...** added to the menu.

Figure 3.3 Data Analysis tool Histogram dialogue box

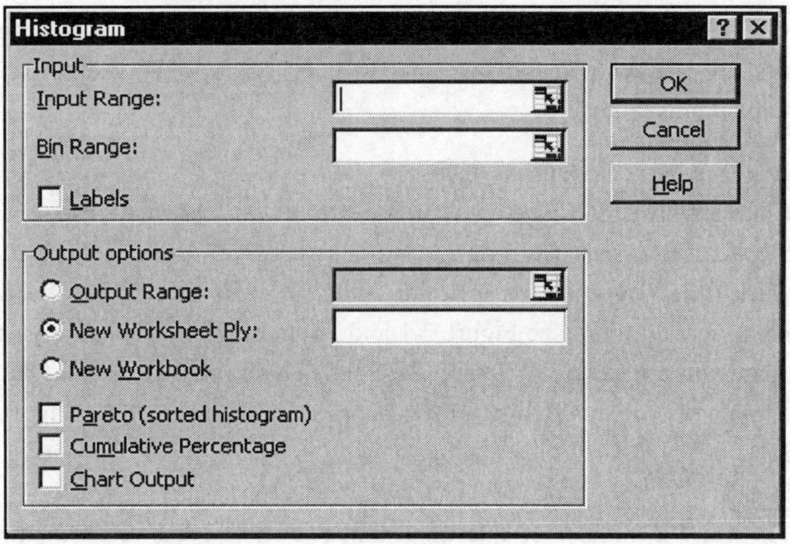

On the other hand, if the frequency distribution table is already given, we can use the icon **ChartWizard** available on the *Standard toolbar* to draw the histogram of this distribution following a 5-step process.

As an alternative to the histogram, a *stem and leaf* display can be used to organise quantitative data and is useful in preliminary data analysis. However, histogram remains the best display to use in formal presentations.

Aim: At the completion of the examples in this Chapter, you should be able to use EXCEL to calculate relative and cumulative frequencies; draw histograms, Ogive curves, frequency polygons, pie charts and line charts; and edit graphs.

3.2 Example 3.1: Drawing histograms from raw data

(Ref: Exercise 2.4, p. 21, Australian Business Statistics, Selvanathan et al., 1994)
The results from a statistics exam are as follows:

75	66	77	66	64	73	91	65	59	86
61	86	61	58	70	77	80	58	94	78
62	79	83	54	52	45	82	48	67	55

Use EXCEL to perform the following tasks.

Task 1 Construct a frequency Histogram for the exam results.

Task 2 Construct an Ogive curve.

Task 3 Save your worksheet on your disk with file name 'A:EX31.XLS' and print the file.

Solution to Example 3.1

First open a new EXCEL workbook. On the worksheet that appears, type the column heading 'Exam Results' in cell A1 and the raw data in cells A2 to A31. Adjust the column width to fit the column heading.

Task 1 ***Constructing a frequency histogram from raw data***

Case 1 ***Class intervals are to be constructed automatically by EXCEL***

Step 1

Click the **Tools** option on the *Menubar* and select the **Data Analysis...** option (see Figure 3.1 and Footnote 1). A Data Analysis menu will appear (see Figure 3.2). Select **Histogram** and click **OK**. A dialogue box for Histogram will appear on the screen (see Figure 3.3).

Step 2

There are a number of questions that need to be answered in the input and output dialogue boxes.

Input Options

Input Range: Enter the Input Range as A2:A31. Either you can manually type **A2:A31** or just browse the mouse over cells A2 to A31. (If you want to use the column title and label for the chart, you should select the Input Range as A1:A31.)

Bin Range: Not applicable in this Case 1, as EXCEL will automatically calculate the class limits.

Labels: Tick this box only if your Input Range includes the column title (i.e. if you have assigned the Input Range above as A1:A31, then you should tick this box).

Output Options

Output can be saved in either of the 3 positions: (1) on the same worksheet[2]; (2) on a different worksheet; or (3) in a new workbook.

Output Range: Click on the little circle against **Output Range** and type the output range[2] (the area where you want the output to be displayed in your current worksheet for option (1) above), **D1**, say. Starting point for the output range is sufficient.

New Worksheet Ply: The output is saved on a new worksheet of the current workbook if you click this option. You will need to type a name in the rectangular space. [Note that the name of the original worksheet is **Sheet 1**.]

New Workbook: The output is saved on a separate workbook if you select the **New Workbook** option.

---

[2] To avoid overwriting the raw data with the output, the Output Range should not overlap the Input Range.

Pareto: To obtain a standard frequency distribution and/or chart, leave this box blank. If you click on the square box against **Pareto**, then the class intervals are sorted according to frequency.

Cumulative Percentage: To construct an Ogive curve (i.e. cumulative relative frequency polygon), tick this box.

Chart Output: To obtain a histogram in addition to the usual frequency table output, tick this box.

Step 3

After completing the input and output options (see Figure 3.4), click **OK**. Now you will see the frequency table and the Histogram (for your data with class intervals selected automatically) displayed starting from cell D1 (see Figure 3.5). [If the labels on the X-axis are not displayed horizontally, to change to horizontal alignment, do the following: Right click the mouse on the X-axis; select **Format Axis...**; on the *Format Axis dialogue box*, select **Alignment**; and under *Orientation*, for *Degrees*, type **0**; and click **OK**.]

Figure 3.4 Histogram dialogue box (class intervals not specified)

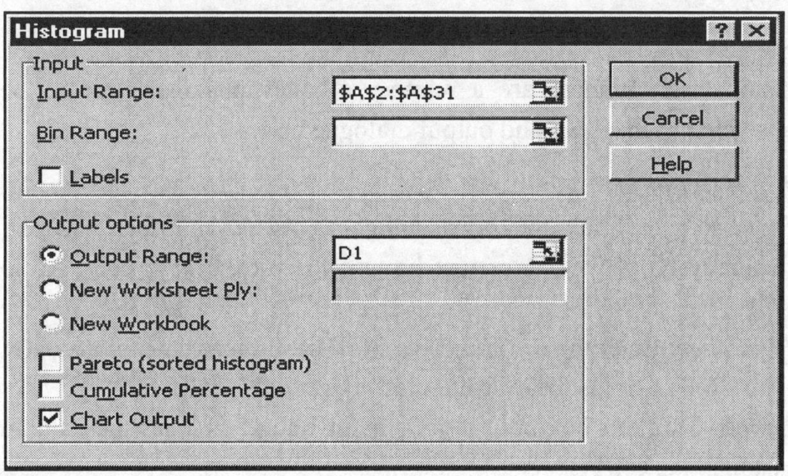

Figure 3.5 Histogram for frequency distribution (class intervals not specified)

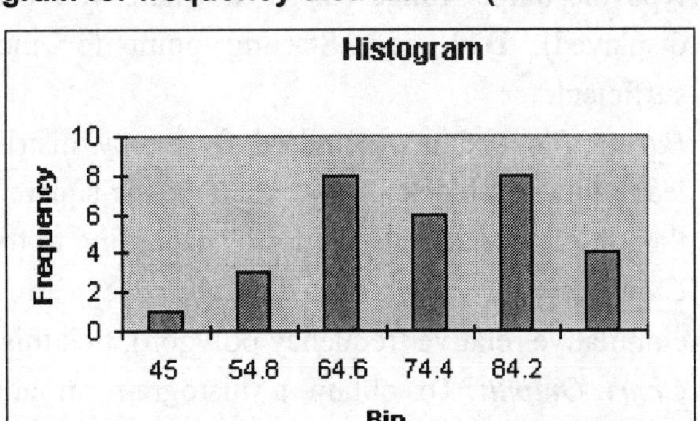

Case 2 ***Class interval limits are to be specified***

After keying in the data in cells A2 to A31, determine the minimum and maximum values of the data. In cell B2, type the label 'Minimum=', and in cell C2, type the formula =**MIN(A2:A31)**. In cell B3, type the label 'Maximum=', and in cell C3, type the formula =**MAX(A2:A31)**. For the exam results, you will find that the minimum value is 45 and the maximum is 94 and there are 30 observations. Suitable class intervals would be 40 to 50, 50 to 60, 60 to 70, 70 to 80, 80 to 90, and 90 to 100. In EXCEL, the upper limits of the class intervals are called **BINS**. In cell B5, type the label 'BINS'. Enter the upper limits of the class intervals **50, 60, …., 100** in cells B6, B7, …., B11, respectively.

Step 1

Click the **Tools** option on the *Menubar* and select the **Data Analysis…** option (see Figure 3.1 and Footnote 1). A Data Analysis menu will appear (see Figure 3.2). Select **Histogram** and click **OK**. A *Histogram dialogue box* will appear on the screen (see Figure 3.3).

Step 2

As in Case 1, there are a number of questions that need to be answered in the input and output dialogue boxes.

Input Options

Input Range: Enter the *Input Range* as **A2:A31**. Either you can manually type A2:A31 or just browse the mouse over cells A2 to A31.

Bin Range: Enter the *Bin Range* as **B6:B11**. Either you can manually type B6:B11 or just browse the mouse over cells B6 to B11.

Labels: Tick this box only if your Input Range includes the column heading, i.e. cell A1.

Output Options

Output Range: Click on the little circle against *Output Range* and type the output range (the area where you want the output to be displayed), **D20**, say. Starting point for the Output Range is sufficient.

Pareto: To obtain a standard frequency distribution and/or chart, leave this box blank. If you click on the square box against **Pareto**, then the class intervals are sorted according to frequency.

Cumulative Percentage: To construct an Ogive curve (i.e. cumulative relative frequency polygon), tick this box.

Chart Output: To obtain a histogram in addition to the usual frequency table output, tick this box.

Step 3

After completing the input and output options (see Figure 3.6), click **OK**. Now you will see the frequency table and the Histogram (for your data in column A with the specified class intervals) displayed starting from cell D20 (see Figure 3.7).

Figure 3.6 Histogram dialogue box (class intervals specified)

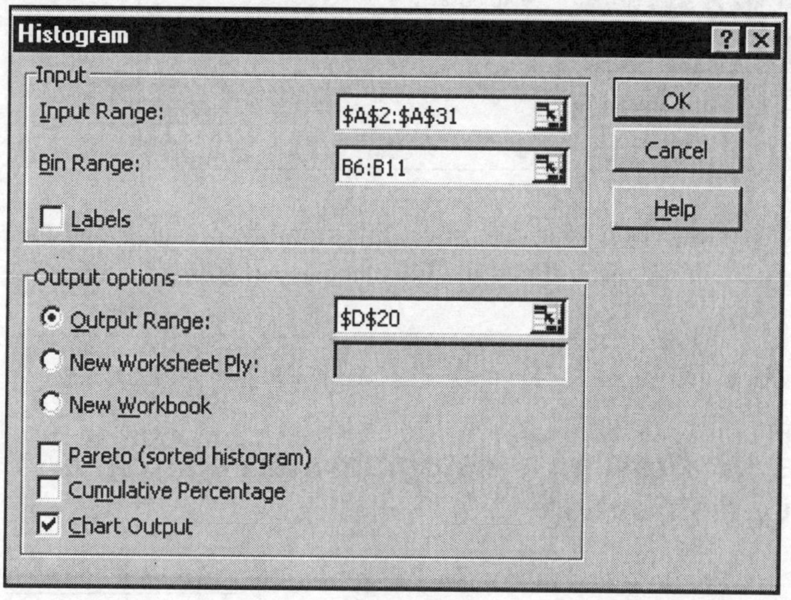

Figure 3.7 Histogram for frequency distribution (class intervals specified)

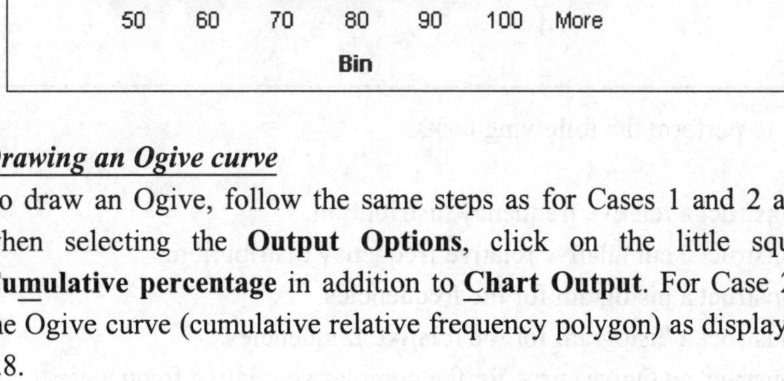

Task 2 ***Drawing an Ogive curve***

To draw an Ogive, follow the same steps as for Cases 1 and 2 above except when selecting the **Output Options**, click on the little square against **Cumulative percentage** in addition to **Chart Output**. For Case 2, this gives the Ogive curve (cumulative relative frequency polygon) as displayed in Figure 3.8.

Figure 3.8 Histogram and Ogive for frequency distribution

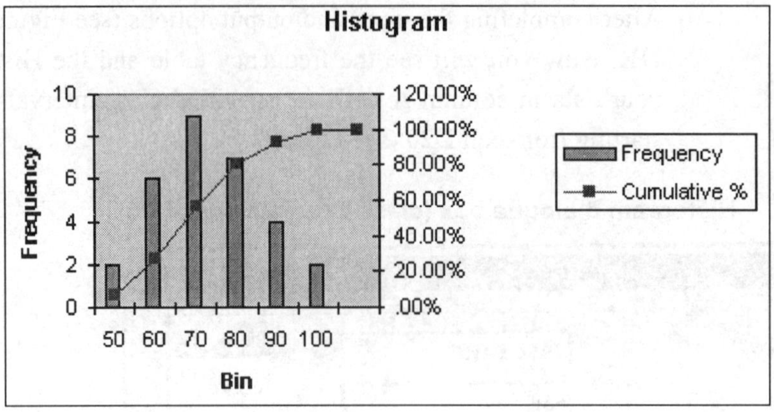

Task 3 To save your worksheet on your disk select **File** from the *Menubar* and then select **Save As** from the menu. Enter the file name as **A:EX31.XLS** and click **OK**.

3.3 Example 3.2: Drawing a histogram starting from a frequency distribution

(Ref: Based on the example illustrated in Section 2.3, pp. 11-18, Australian Business Statistics, Selvanathan et al., 1994)

The following is the frequency distribution of the duration of interstate telephone calls of a firm during a week.

Frequency Distribution of Telephone-Call Durations

Class Intervals	Frequency
2 to 5	3
5 to 8	6
8 to 11	8
11 to 14	7
14 to 17	4
17 to 20	2

Use EXCEL to perform the following tasks.

Task 1 Construct a relative frequency distribution.
Task 2 Construct a cumulative relative frequency distribution.
Task 3 Construct a histogram for the frequencies.
Task 4 Construct a histogram for the relative frequencies.
Task 5 Construct an Ogive curve for the cumulative relative frequencies.
Task 6 Draw a three-dimensional frequency polygon.

Task 7 Centre the heading 'Frequency Distribution of Telephone-Call Durations' across the columns. Bold the heading and the column headings.

Task 8 Save your worksheet with file name 'A:EX32.XLS' on your disk and print the file.

Solution to Example 3.2

First open an EXCEL worksheet window. On the worksheet, type 'Frequency Distribution of Telephone Call Durations' in cell A1, and type the column headings 'Class Intervals' in cell A2 and 'Absolute Frequencies' in cell B2. Then, type 'Relative Frequencies' in cell C2 and 'Cumulative Frequencies' in cell D2. Adjust the column widths to fit the column headings. Use columns A and B to type the given class limits and the corresponding frequencies (see Figure 3.9).

Task 1 ***Constructing a relative frequency distribution***

To calculate the relative frequencies, we divide each frequency by the total frequency. First, we compute the total frequency. To obtain the total frequency, highlight cells B3 to B8 and click on the icon Σ **Autosum** on the *Standard toolbar*. The sum, which is 30, appears in cell B9. Notice the formula which appears on the *Formula bar* when you activate cell B9. Type the label 'Total' in cell A9.

To calculate the first relative frequency in cell C3, type the formula **=B3/$B$9**. Note that instead of typing B9, we type $B$9 because when we copy (see Chapter 2) the formula into the other cells of that column, EXCEL will use the same value in cell B9 as the denominator for each cell while changing the numerator from B3 to B4, B5, and so on. Now, copy the formula in cell C3 to cells C4 through to C8 to obtain the other relative frequencies using the pointer +. As a check, sum all the relative frequencies in cell C3 to C8 which should add-up to 1. To do this, after highlighting cells C3 through to C8, click on the icon Σ **Autosum**, the sum 1 should appear in cell C9.

Task 2 ***Constructing a cumulative frequency distribution***

As the first cumulative relative frequency is the same as the first relative frequency, type =**C3** in cell D3. To obtain the second cumulative frequency in cell D4, add the corresponding relative frequency in cell C4 to the previous cumulative relative frequency in cell D3. That is, enter the formula **=C4+D3** in cell D4. To calculate the remaining cumulative frequencies in cells D5 to D8, copy this formula in cell D4 onto the remaining cells D5 to D8 using the pointer +. You will notice that when you copy the formula in D4 onto D5, the formula will change automatically from **=C4+D3** to **=C5+D4**. As a check, you should verify that the cumulative relative frequency in the last cell D8 is 1.00. See Figure 3.9 for the output.

Figure 3.9 Frequency distribution for telephone call durations

	A	B	C	D
1	Frequency Distribution of Telephone Call Durations			
2	Class Intervals	Absolute Frequencies	Relative Frequencies	Cumulative Frequencies
3	2 to 5	3	0.10	0.10
4	5 to 8	6	0.20	0.30
5	8 to 11	8	0.27	0.57
6	11 to 14	7	0.23	0.80
7	14 to 17	4	0.13	0.93
8	17 to 20	2	0.07	1.00
9	Total	30	1.00	

Task 3 ***Constructing a histogram for the frequencies***

To draw the histogram for the frequencies in column B corresponding to the class intervals in column A, first highlight the class intervals and the frequencies in cells A3 to B8. [In this example, as the two columns are adjacent to each other, we highlighted A3 to B8 and the whole range is selected automatically. However, if the columns are apart, say in columns A and C, first highlight the entries in column A first and then while holding the **CTRL** key down highlight the entries in column C. This will highlight the entries in both columns A and C.]

Next click on the chart symbol icon **ChartWizard** on the *Standard toolbar*. Click the mouse on the worksheet area where you would like the histogram to be displayed. You will see a **ChartWizard - Step 1 of 4** window on the screen (see Figure 3.10). On this window, select *Chart type:* **Column** and select a *Chart sub-type:* (say for example, the first selection in row 1). Now you will see both the selected chart type and chart sub-type highlighted (see Figure 3.10).

Now click on **Next>** to go to Step 2. On the Step 2 window, the **Data range:** rectangle will have entries as the highlighted cells A3 to B8 appearing as **=Sheet1!$A$3:$B$8**. *Alternatively*, instead of highlighting the data range before-hand, if you already know the cell location of the data (i.e., A3-A8 and B3-B8), you can simply type the range as either **A3:A8,B3:B8** or **A3:B8** (since the data are in adjacent columns), or browse the mouse over cells A3 to B8. For *Series in:* click the circle against **Columns** (as our data are in columns). In this step, you will also notice a sample chart display of your data (see Figure 3.11). This concludes Step 2.

Now click on **Next>** to go to Step 3. In Step 3, you will see the sample chart together with options to include chart title, titles for *X* and *Y*-axes. Click the mouse and type a *Chart title* in the space provided. For example, type 'Histogram: Telephone Call Durations'. Then press the **TAB** key or click the mouse on the space for *Category (X) axis:*, type 'Class Intervals', and for *Value (Y) axis:*, type 'Absolute Frequencies' (see Figure 3.12). This concludes Step 3.

Figure 3.10 Chartwizard Chart Types

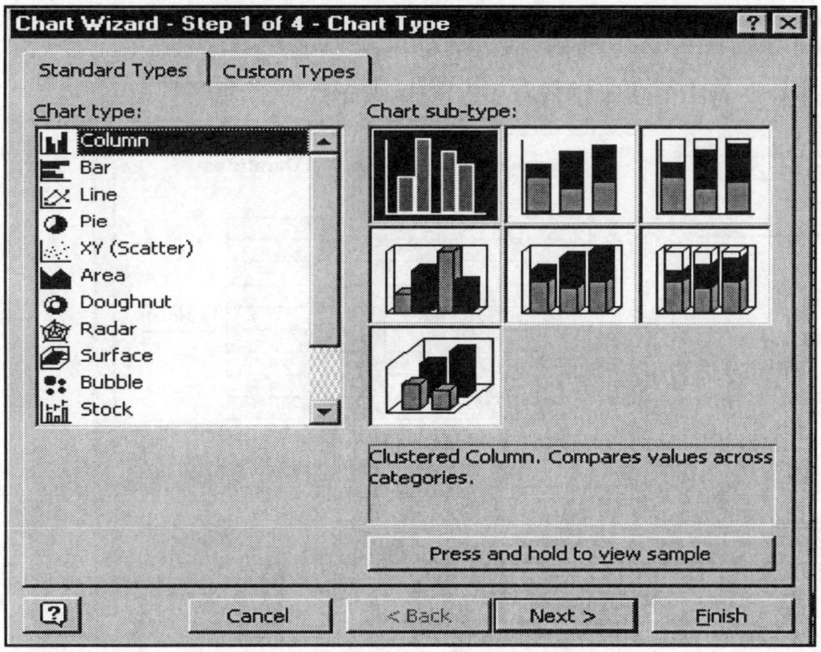

Figure 3.11 Data range for histogram

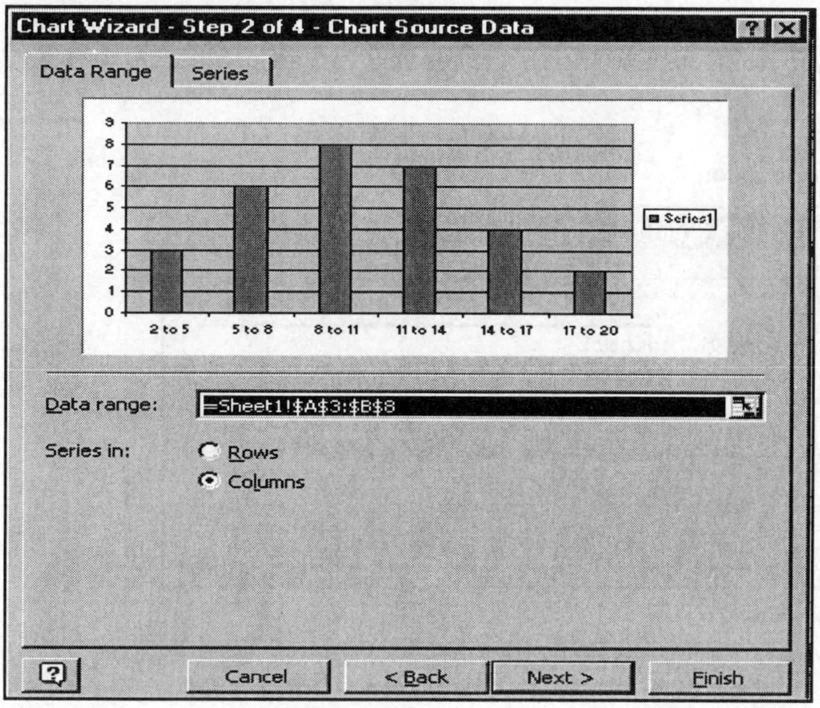

Figure 3.12 Chart options

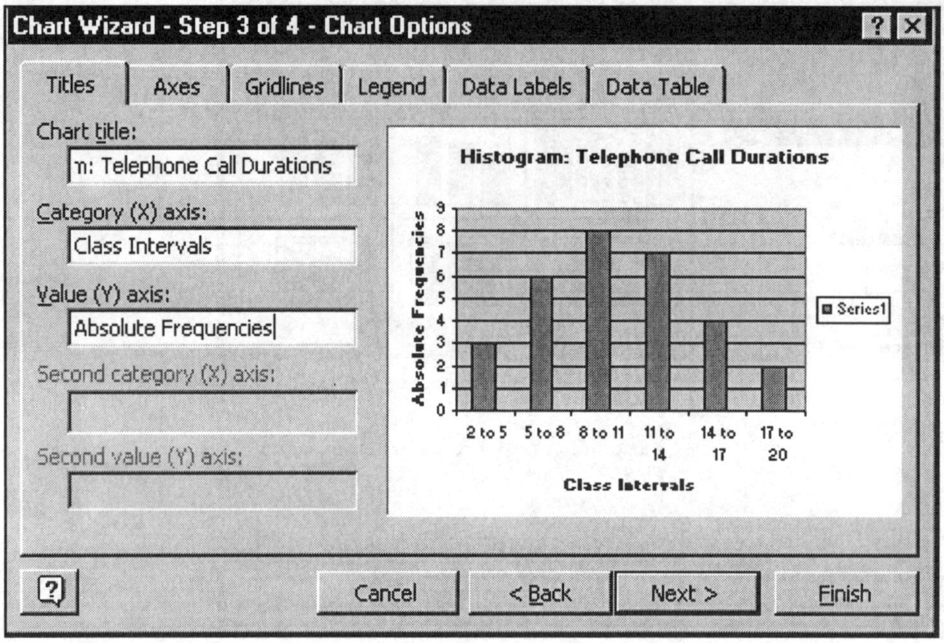

Now click on **Next>** to go to Step 4. In Step 4, if you want the graph to be included on the same worksheet as your data, click the circle against *As object in:* and type the sheet name (e.g., Sheet 1 if your data are in Sheet 1), see Figure 3.13.

Figure 3.13 Chart location

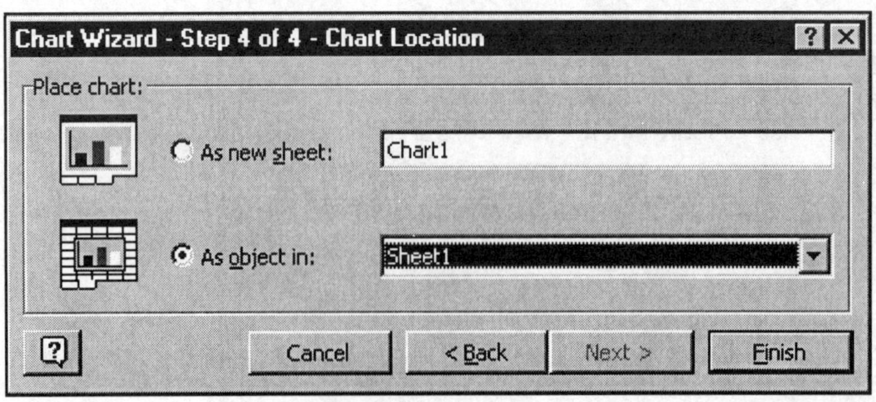

Finally click **Finish** of Step 4. Now you will see the histogram displayed as in Figure 3.14.

Figure 3.14 Frequency histogram for telephone call durations

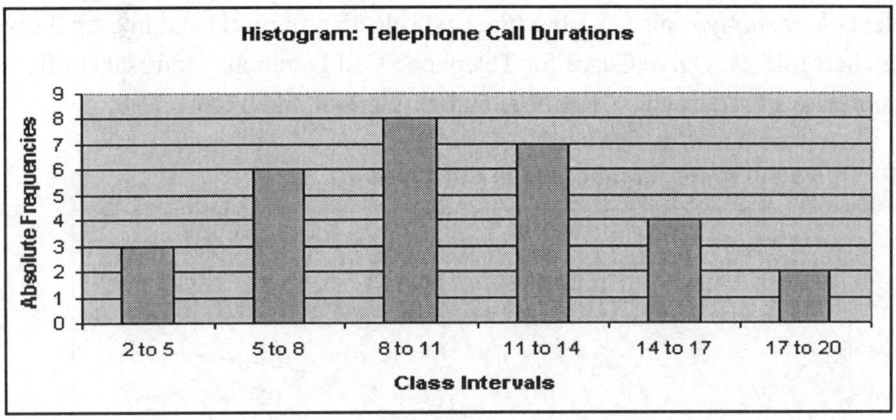

Task 4 *Constructing a histogram for the relative frequencies*

To draw a histogram of relative frequencies, we need the information in columns A and C. To do this, highlight A3 to A8 first, and then while holding the CTRL key down highlight C3 to C8. Now you will see the entries in A3 to A8 and C3 to C8 highlighted. Then click on the icon **ChartWizard** on the *Standard toolbar* and then click on the worksheet area where you would like the chart to be displayed. Step 1 is as in Task 3. In Step 2, the range **=$A$3:$A$8,$C$3:$C$8** will appear. *Alternatively*, rather than highlighting the data range beforehand, you can also simply enter the range manually as **A3:A8;C3:C8** in the space provided for the *Range*. When done, click **Next>** and proceed as in Steps 3 and 4 of Task 3 (except that the *Y*-axis title should be typed as 'Relative Frequencies' in Step 3). When completed, the relative frequency histogram will look like Figure 3.15.

Figure 3.15 Relative frequency histogram for telephone call durations

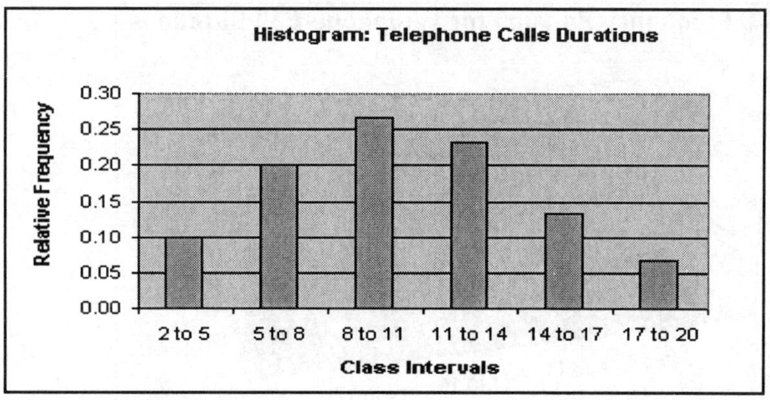

Task 5 *Constructing an Ogive curve*

To draw the Ogive curve for the cumulative relative frequencies, follow the same steps given in Task 4 and use Steps 1 to 4 of **ChartWizard** to draw the

Ogive curve. The difference here is that, in Step 1, select *chart type* **Line** and select *chart sub-type* also as **Line** (the first selection in row1) and in Step 3 give the chart title as 'Ogive Curve for Telephone Call Durations' and *Y*-axis title as 'Cumulative Frequencies'. The final output will look like Figure 3.16.

Figure 3.16 Ogive curve for telephone call durations

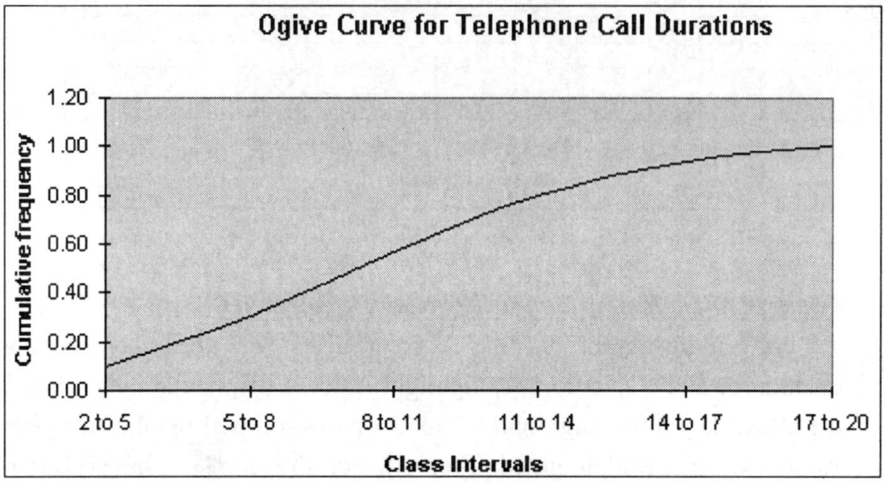

Task 6 ***Drawing a 3-dimensional frequency polygon***

To draw a 3 dimensional frequency polygon for the frequencies in Column B as before, follow the same instructions in Task 3 and follow Steps 1 to 5 of the **ChartWizard**. In Step 1, select *Chart type:* **Line** and *Chart sub-type:* **3-D Line** and follow Task 3. In Step 3, for the chart title, type '3-D Frequency Polygon for Telephone Call Durations' and for the *Y*-axis title type 'Relative Frequency'. The final output will look like Figure 3.17.

Figure 3.17 3-Dimensional frequency polygon for telephone call durations

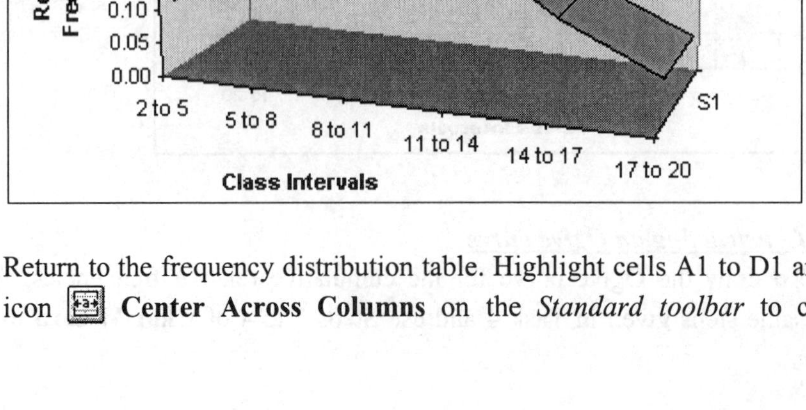

Task 7 Return to the frequency distribution table. Highlight cells A1 to D1 and use the icon [icon] **Center Across Columns** on the *Standard toolbar* to centre the

heading. Highlight the heading and column headings in cells A1 to D2 and bold them using the icon **B** **Bold** on the *Standard toolbar*.

Task 8 To save your work on your disk select **File** from the *Menubar* and then select **Save As** from the menu. Enter the file name **A:EX32.XLS** and click **OK**.

3.4 Editing a Chart

Shifting the graph to another location

To shift the graph to another location, move the pointer ✥ to anywhere inside the rectangular border of the graph and see the pointer changing to ↖. Click the mouse and drag the graph, and release at the desired location.

Adjusting the height and the width of the graph

When you click the mouse inside the rectangular graph area, you will notice a rectangle with 8 selection handles like this ▭ to appear. Now, if you want to increase/decrease the height of this rectangle area, simply move the pointer ↖ to one of the little black selection handles (■) on the middle of the horizontal boundaries. The pointer will take a ↕ shape. Then drag this up or down to change the height of the rectangle. Similarly, to increase/decrease the width of the rectangle area, use the selection handles on the middle of the vertical boundaries. Use the selection handles on the corners to change both the height and weight of the graph.

Editing within the graph area

To edit text within the rectangular area, click the mouse on the text. You will notice a thicker border appearing around the text (see Figure 3.18). This means that you are now in the **Edit** mode and can make changes inside the text area. For example, if you want to change the *X*-axis title in Figure 3.15 from 'Class Intervals' to 'Classes', click the mouse on 'Class Intervals' and replace it with **Classes** by typing that word. When done, click the cursor anywhere outside the rectangular border (see Figure 3.18). Now, the pointer ✥ reappears on the active cell.

Removing the gaps between the bars of a histogram chart

As can be seen from the histograms obtained in Figures 3.5 (Case 1) and 3.7 (Case 2) in Section 3.2, there are gaps between the rectangular bars, whereas in a standard histogram, there are no gaps. To remove the gaps, first click the mouse on one of the rectangular bars. You will notice little black squares appearing in the middle of the rectangular bars (see Figure 3.19). Point the mouse on one of these black squares and click the **right** button of the mouse. A menu will appear (see Figure 3.19). From this menu, select **Format Data Series…** A dialogue box for *Format Data Series* will now appear. Click on **Options** to see another dialogue box (see Figure 3.20) and in this dialogue box change the **Gap Width** to zero using the down arrow. This will remove the gaps between the rectangular bars. Figure 3.21 displays the histogram in Figure 3.7 with the gaps removed.

Figure 3.18 Histogram with edited X-axis title

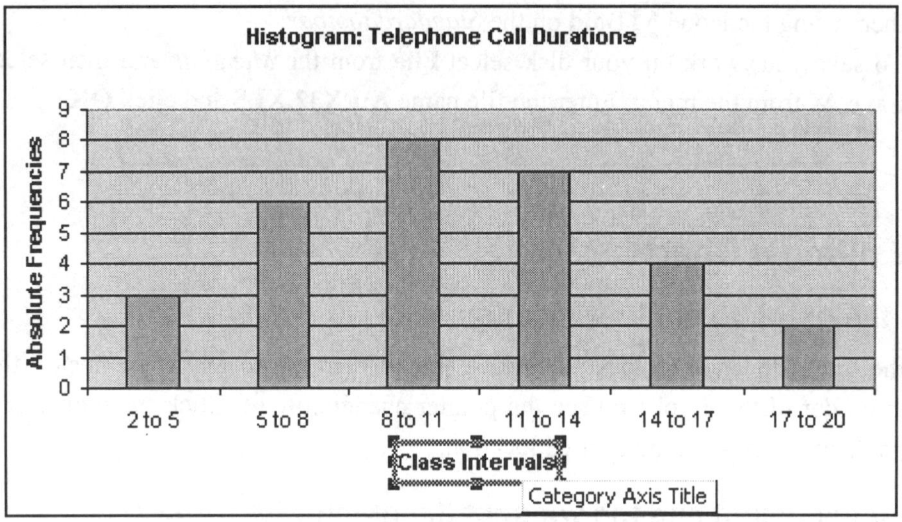

Figure 3.19 Formatting a Histogram menu

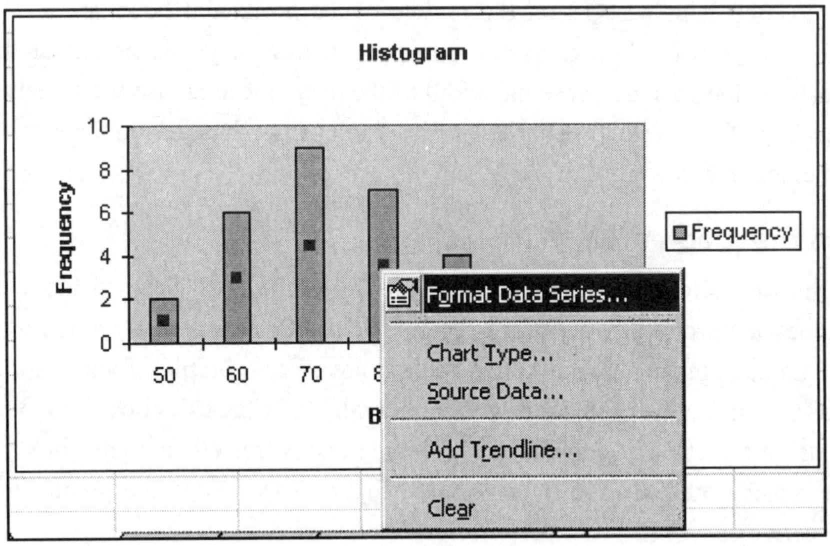

Formatting the axes of the chart

As can be seen, the histogram obtained in Figure 3.21 has the scales on the *Y*-axis as 0, 2, 4, 6, 8, 10. That is, the *Y*-axis scales have an increment of 2. Suppose we want to change the increment to 1, i.e., change the scales on the *Y*-axis to 0,1,2, …,9,10. First, point the mouse on the *Y*-axis and click the **right** button of the mouse. A menu will appear close to the *Y*-axis (see Figure 3.22). From the menu select the **Format axis...** option. From the dialogue box appearing on the screen, select **Scale**, delete the tick against **Major Unit** and change the **Major unit** from 2 to 1 (see Figure 3.23). Click on **OK**. Now the scale on the Y-axis will be 0, 1, 2, 3, …, 10 (see Figure 3.24). Note that if the height of the Y-axis is too small, you will not see the change in the scales. This can be fixed by increasing the height of the Y-axis as discussed earlier.

Figure 3.20 Editing the gap dialogue box

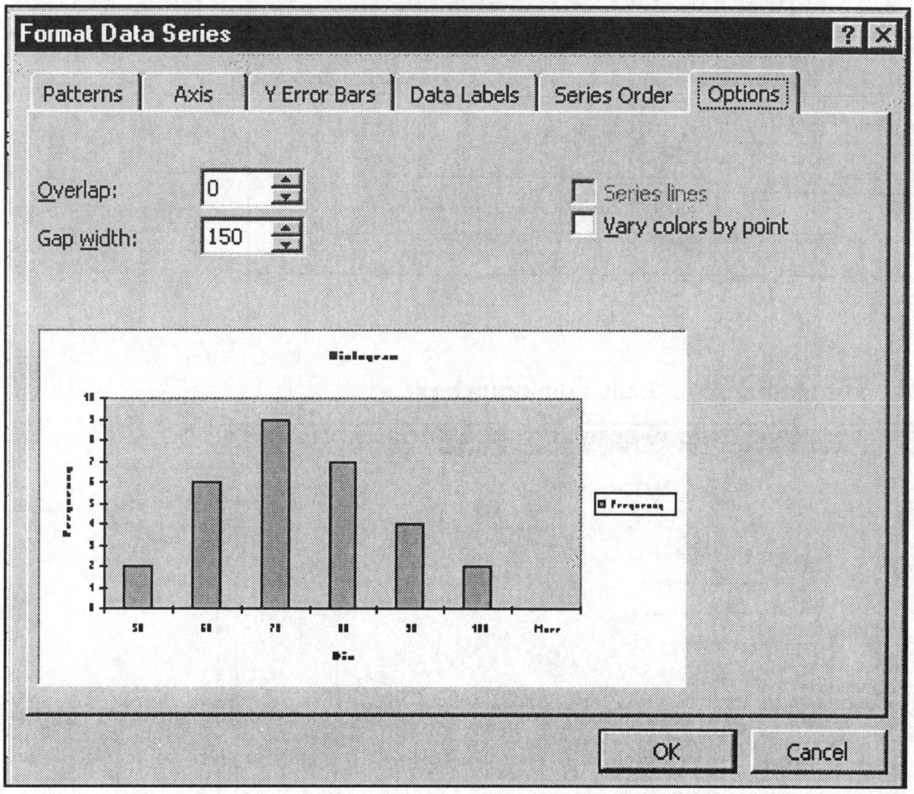

Figure 3.21 Edited histogram

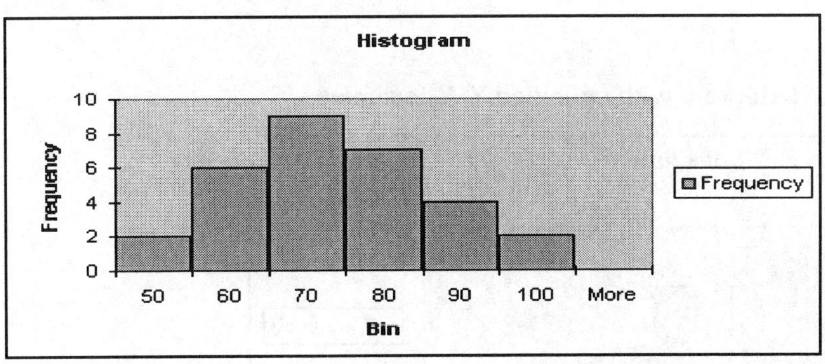

Figure 3.22 Formatting Axis Scales menu

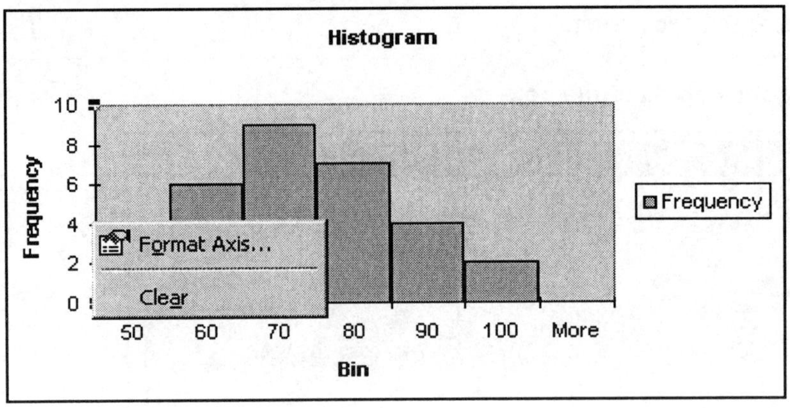

Figure 3.23 Formatting Axis Scales dialogue box

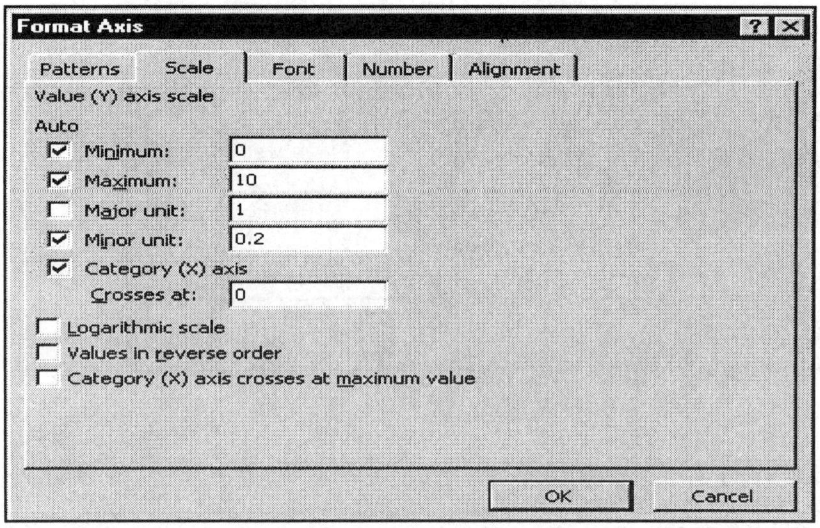

Figure 3.24 Edited histogram with specified Y-Axis scales

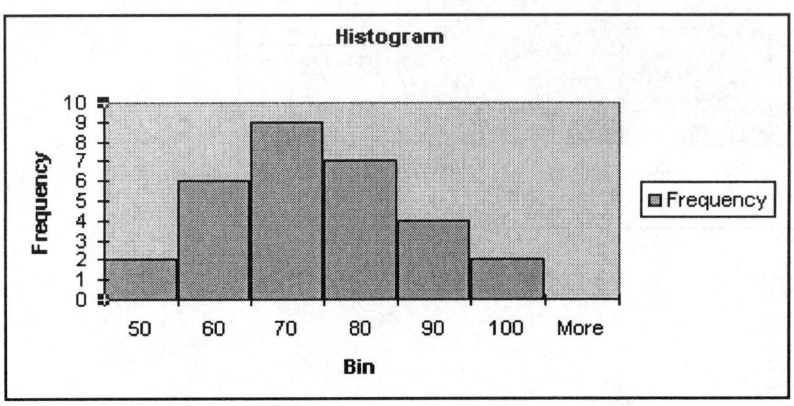

3.5 Pie charts and bar charts

Pie charts and bar charts are primarily used for graphical presentation of qualitative data. In some instances, they are also used to present frequency distributions of quantitative data.

A *Pie chart* is a circle subdivided into a number of slices that represent the various categories in a data set. Pie charts are effective whenever the objective is to display the components of a whole entity indicating their relative sizes.

A *Bar chart* represents the frequency (or relative frequency) of each of the categories (of a data set) as a bar with height proportional to the frequency (or relative frequency) rising vertically (or horizontally) from the horizontal (or vertical) axis. To construct a pie chart or bar chart using EXCEL, we use the icon **ChartWizard** available on the *Standard toolbar* and by selecting the relevant graph options.

3.6 Example 3.3: Constructing pie charts and bar charts

The following table presents the market shares of new passenger vehicle registrations in Australia in October 1997 and October 1998.

Percentage market shares by manufacturer of new passenger vehicle registrations, Australia, October 1997 and October 1998

Manufacturer	Oct 1997	Oct 1998
Ford	16%	17%
Holden	18%	18%
Hyundai	9%	7 %
Mazda	4%	3%
Mitsubishi	12%	10%
Nissan	4%	6%
Toyota	15%	18%
Other	22%	21%

Task 1 Use a pie chart and a bar chart to show the October 1998 percentage market shares of new vehicle registrations, broken down by manufacturer.

Task 2 Use a bar chart to compare the October 1997 and October 1998 market shares.

Task 3 Save your worksheet on your disk with file name 'A:EX33.XLS' and print your output.

Solution to Example 3.3

Task 1 Type the heading 'Market shares by manufacturer of new passenger vehicle registrations, Australia, October 1998' in cell A1 and the column headings, 'Manufacturer', 'October 1997' and 'October 1998' in cells A2, B2 and C2. Type the data in cells A3 to C10.

Now we draw a pie chart for the October 1998 percentages in Column C broken down by manufacturer in Column A (i.e., use A3 to A10 and C3 to C10). As in the last two examples, we follow Steps 1-4 of **ChartWizard**. The difference here is that in Step 1, select *Chart type:* Pie and *Chart sub-type:* 3-D visual effect (second on the top row). In Step 2, type the *Data range:* **A3:A10; C3:C10**, and for *Series in:* click the circle for **Columns**. In Step 3, for Chart Options:, click on **Titles** and type *Chart title:*, 'Market shares by manufacturer of new passenger vehicle registrations, Australia, October 1998'. Click on **Legend** and then remove the tick for **Show Legend** by clicking on the tick, and click on *Data Labels* and tick the circle for **Show label and percent**. In Step 4, tick on *As object in:* and type Sheet 1. The resulting 3-D Pie chart with appropriate chart title included will look like Figure 3.25.

Figure 3.25 3-Dimensional Pie chart of market shares by manufacturer

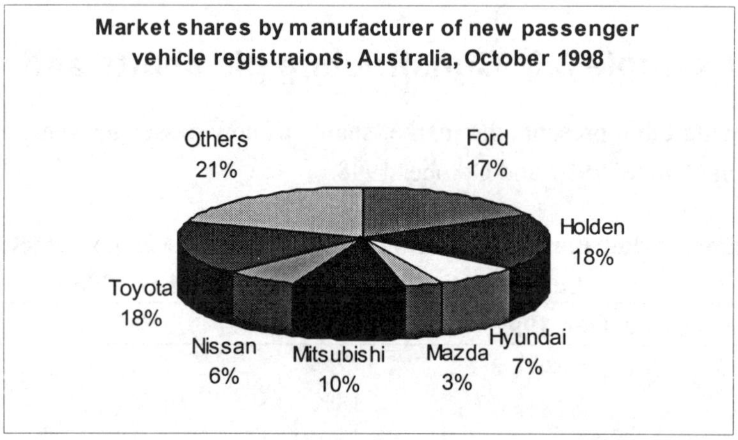

To draw a bar chart for the same data as for the pie chart, we follow Steps 1-4 of **ChartWizard**. In Step 1, select **Bar** for the *Chart type* and **Clustered Bar** (first selection in row 2) for the *Chart sub-type*. In Step 3, click on **Titles** and type the chart title, and then click on **Data Labels** and click the circle for **None** (see Figure 3.10). Everything else is the same as earlier. The resulting bar chart will look like Figure 3.26.

Task 2 To draw a bar chart to compare the October 1997 and October 1998 percentages of market shares by manufacturer, again we follow Steps 1-4 of **ChartWizard**. In Step 1, select *Chart type:* **Column** and *Chart sub-type:* **Clustered column** (first selection in row 1). Click **Next>**. In Step 2, enter the *Data range:* as **A2:C10** and for *Series in:* click the circle for **Columns** and click **Next>**. In Step 3, click on *Title* and type chart title, 'Market shares by manufacturer of new passenger vehicle registrations, Australia, October 1997 and October 1998', and for X-axis, type 'Manufacturer' and for Y-axis, type 'Percentage'. Step 4 is the same as earlier. The resulting bar chart will look like Figure 3.27.

Figure 3.26 Bar chart of market shares by manufacturer

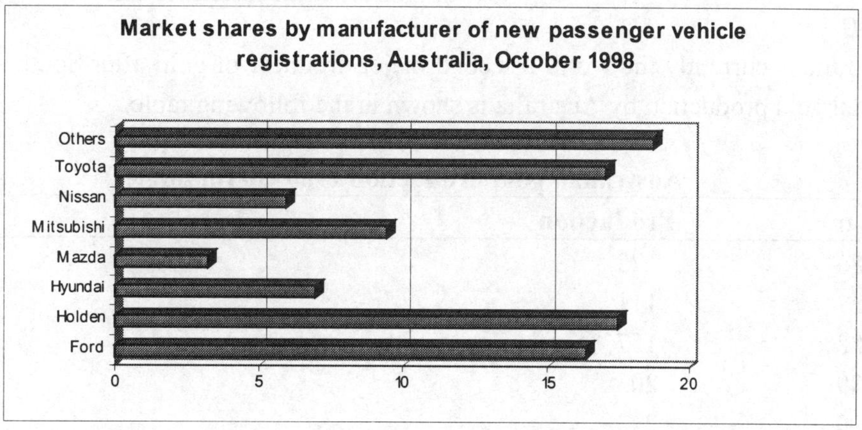

Figure 3.27 Bar chart of market shares, October 1997 and October 1998

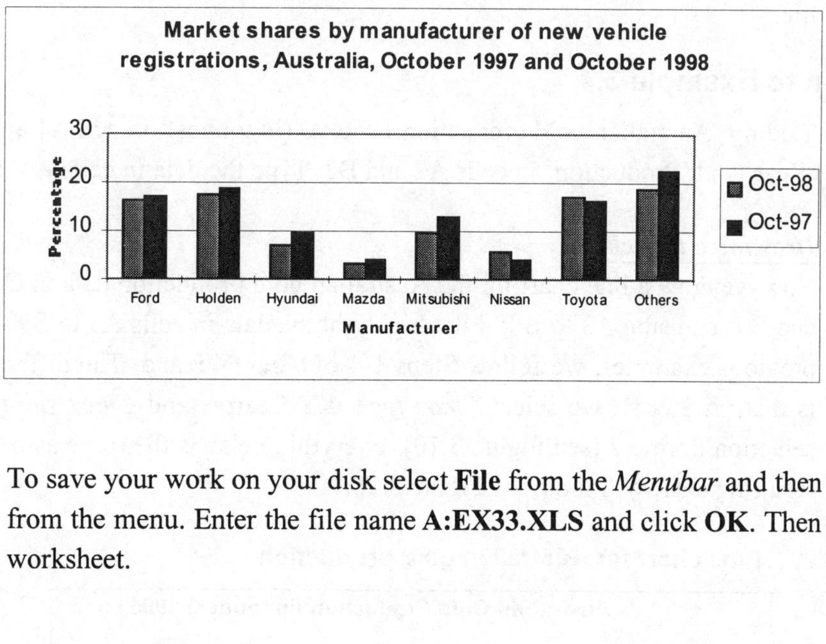

Task 3 To save your work on your disk select **File** from the *Menubar* and then **Save As** from the menu. Enter the file name **A:EX33.XLS** and click **OK**. Then print the worksheet.

3.7 Line charts

A **line chart** is obtained by plotting the frequency of a category above the point on the horizontal axis representing that category, and then joining these points with straight lines. Normally line charts are used when the categories are points in time, and the resulting line chart is called a **time-series chart**. To construct a line chart using EXCEL, we use the icon **ChartWizard** available on the *Standard toolbar* by selecting the relevant graph options.

3.8 Example 3.4: Drawing line charts

(Ref: Based on Exercise 2.15, p.30-31, Australian Business Statistics, Selvanathan et al., 1994)

Australia is currently the world's second major producer of gold after South Africa. The annual gold production by Australia is shown in the following table.

Australian gold production 1986–90 (in tonnes)

Year	Production
1986	75
1987	111
1988	157
1989	203
1990	243

Task 1 Use a line chart to display the data.

Task 2 Save your worksheet on your disk with file name 'A:EX34.XLS' and print the file.

Solution to Example 3.4

Type the heading 'Australian gold production 1986-90 (in tonnes)' in cell A1 and column headings 'Year' and 'Production' in cells A2 and B2. Type the data in cells A3 to B7.

Task 1 ***Drawing a line chart***

Now we draw a line chart for the Australian gold production data in Columns A and B (i.e., using A3 to B8). First highlight the data in cells A3 to B9. As in the previous examples, we follow Steps 1-4 of **ChartWizard**. The difference here is that, in Step 1, we select *Chart type:* **XY Scatter** and *Chart sub-type:* first selection in row 2 (see Figure 3.10). Everything else is the same as before. The resulting chart will look like Figure 3.28.

Figure 3.28 Line chart for Australian gold production

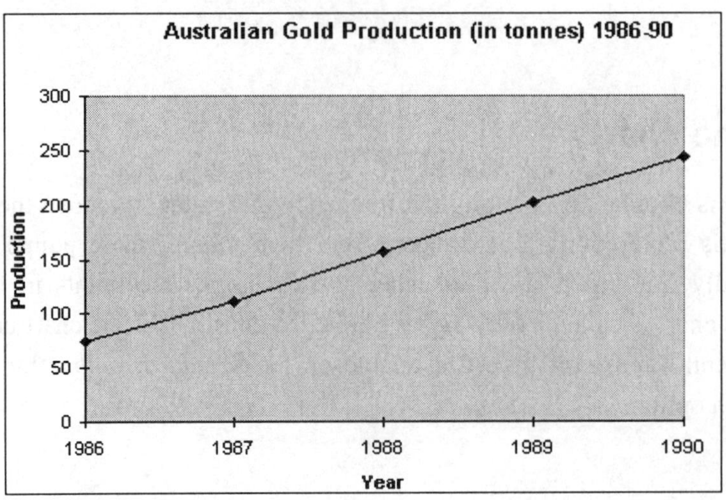

Task 2 Save your worksheet on your disk with file name 'A:EX34.XLS' and print your output.

3.9 Exercises to Chapter 3

(The following exercises are based on exercises from Chapter 2 of Selvanathan et al., Australian Business Statistics, 1994)

Exercise 3.1

The percentage distribution of Australian family incomes for 1990 is shown in the following table:

Percentage distribution of mean gross weekly Australian family income for 1990

Class limits ($)	Percentage
0 up to 200	8.8
200 up to 400	21.0
400 up to 600	23.9
600 up to 800	17.2
800 and over	29.1

Use EXCEL to perform the following tasks.

Task 1 Construct a relative frequency distribution.
Task 2 Construct a cumulative relative frequency distribution.
Task 3 Construct a histogram for the frequencies.
Task 4 Construct a histogram for the relative frequencies.
Task 5 Construct an Ogive curve for the cumulative relative frequencies.
Task 6 Draw a three-dimensional frequency polygon.
Task 7 Save the worksheet on your disk as 'A:INCOME.XLS' and print the file.

Exercise 3.2

A real-estate company in a wealthy suburb of Adelaide wishes to investigate the distribution of prices of homes sold during the past year. The following sample of prices (in $1,000s) was collected:

274	429	229	435	260
222	292	419	242	202
235	215	390	359	409
375	209	265	440	365
319	338	414	249	279

Use EXCEL to perform the following tasks.

Task 1 Construct a frequency distribution for the prices, using five class intervals and the value 200 as the lower limit of the first class.

Task 2 Construct a relative frequency histogram for the prices.

Task 3 Construct an Ogive for the house prices.

Task 4 Draw a frequency polygon for the prices.

Task 5 Save the worksheet on your disk as 'A:HOME.XLS' and print the file.

Exercise 3.3

Asian countries are becoming Australia's most important trading partners as evidenced by the following table showing the average percentage distribution of Australian exports to and imports from Asian countries during 1989-1991.

Country	Exports 1989-1991	Imports 1989-1991
Japan	27%	19%
ASEAN[a]	11%	6%
Korea	6%	3%
China	3%	3%
Hong Kong	3%	2%
Taiwan	4%	4%
Other Asia[b]	2%	1%
Other	44%	62%

Source: *Imports, ABS Catalogue No. 5433.0. The year 1989, for example, refers to the financial year 1988/89.*

a Brunei, Indonesia, Malaysia, Philippines, Singapore and Thailand.

b Bangladesh, India, Pakistan, Sri Lanka, Vietnam.

Task 1 Use a graphical technique to depict the information given in the table.

Task 2 Explain why you selected the particular technique you used in Task 1.

Task 3 Save the worksheet on your disk as 'A:TRADE.XLS' and print the file.

Exercise 3.4

Nestle S.A., the international food giant, had worldwide sales in 1987 of $23 billion. The percentage distribution of sales according to geographic region is shown in the table below.

Task 1 Show how this information can be conveyed to shareholders, using a graphical technique.

Task 2 Explain why you selected the particular technique you used in Task 1.

Task 3 Save the worksheet on your disk as 'A:NESTLE.XLS' and print the file.

1987 Sales for Nestle

Region	Sales
Australia, New Zealand and Pacific Islands	2%
Europe	43%
North America	29%
Asia	13%
Latin America and Caribbean	10%
Africa	3%

Source: *New York Times, 1 January 1989.*

Exercise 3.5

The following table presents the median prices of established houses in Sydney during the period 1984 to 1992.

Median prices of established houses in Sydney (12 months ending September)

Year	Price
1984	86,000
1986	100,600
1988	161,300
1990	179,000
1991	178,600
1992	184,100

Source: *Real Estate Institute of Australia, 1994.*

Task 1 Use a graphical technique to depict the information in the accompanying table.
Task 2 Explain why you selected the particular technique you used in Task 1.
Task 3 Save the worksheet on your disk as 'A:HOUSE.XLS' and print the file.

Chapter 4

Numerical Descriptive Measures: *Summary Statistics*

While graphical descriptive methods provide a general idea about how the data are distributed between the two extreme values, numerical descriptive measures provides numerical measurements to describe the central location of the data and the dispersion of the data relative to the central location.

In EXCEL, there are two approaches available to calculate these summary measures. Either we can use specific functions to calculate individual summary measures or we can use a single function to calculate a list of summary measures. In the first approach, we select the icon *fx* **Paste Function** from the *Standard toolbar* and then select the *Function Category* **All** from the *Paste Function dialogue box* (see Figure 4.1), and then select the *Function name* of the required summary statistic.

Figure 4.1 Statistical Function categories

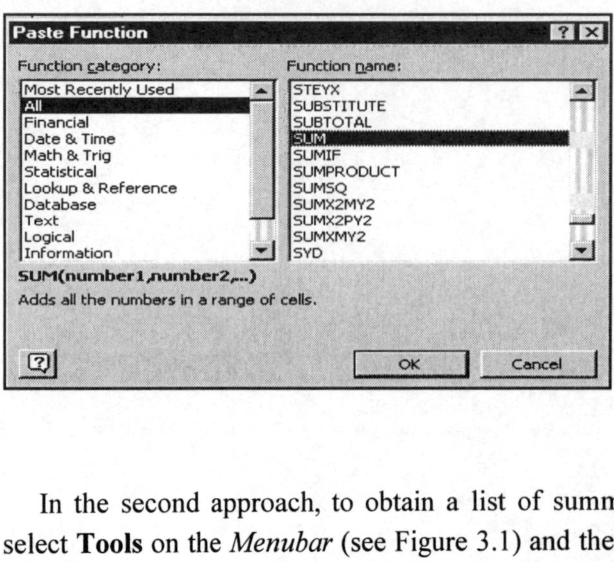

In the second approach, to obtain a list of summary statistics, as in Chapter 3, we select **Tools** on the *Menubar* (see Figure 3.1) and then select the option **Data Analysis...** from the available options. Now from the *Data Analysis dialogue box* (see Figure 4.2)

select the option **Descriptive Statistics**, and then select the **Summary Statistics** option from the *Descriptive Statistics menu* (see Figure 4.3).

Figure 4.2 Data Analysis dialogue box

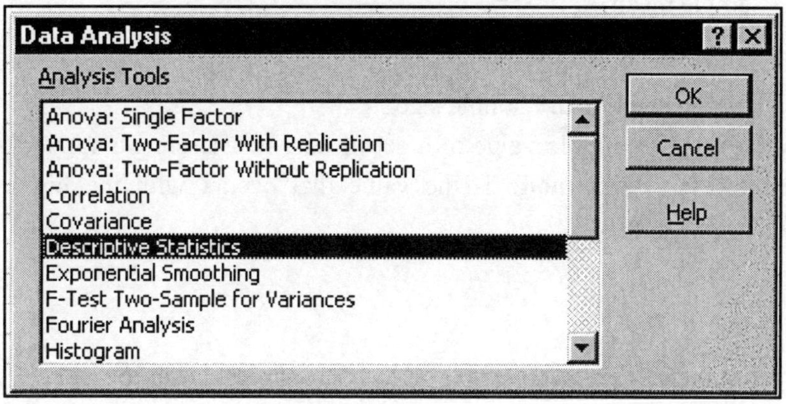

Figure 4.3 Descriptive Statistics dialogue box

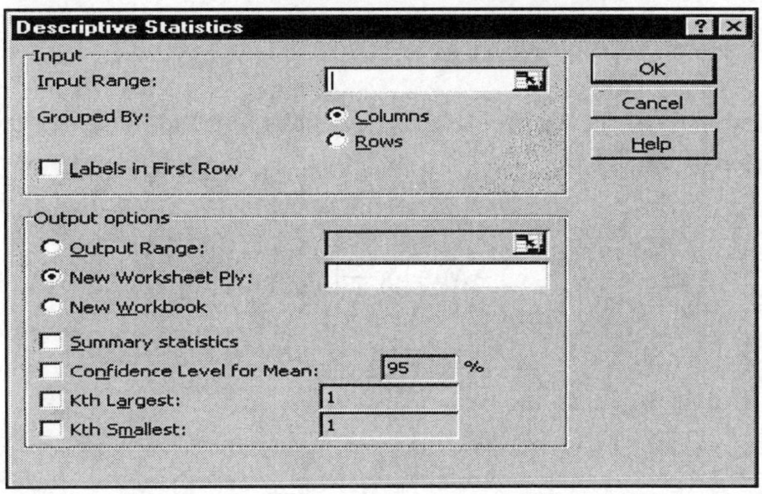

4.1 Measures of central location

Of the various types of measures of central location, the most commonly known are the (arithmetic) mean, median and mode. The most popular and the most useful measure of these three is the **mean** (or the average) which is defined as

$$\text{Population mean } \mu \quad = \quad \frac{\sum_{i=1}^{N} x_i}{N}$$

and

$$\text{Sample mean } \overline{X} = \frac{\sum_{i=1}^{n} x_i}{n}$$

where N is the population size and n is the sample size.

The **median** is defined as the middle value of a set of data measurements ranked in ascending or descending order. The **mode** is the value that occurs with the highest frequency in a data set.

4.2 Measures of dispersion

The dispersion or variability of the data is measured by numerical measures such as range, variance, standard deviation, coefficient of variation, percentiles and quartiles. **Range** is measured as

$$\text{Range} = \text{Maximum value - Minimum value.}$$

The variance is measured as the average mean deviations squared. That is, the **variance for a population** is defined as

$$\sigma^2 = \frac{1}{N}\sum_{i=1}^{N}(x_i - \mu)^2 \text{ or } \frac{1}{N}\left[\sum_{i=1}^{N} x_i^2 - N\mu^2\right] \text{ or } \frac{1}{N}\left[\sum x_i^2 - \frac{\left(\sum x_i\right)^2}{N}\right]$$

and the **variance for a sample** is defined as

$$s^2 = \frac{1}{n-1}\sum_{i=1}^{n}(x_i - \overline{X})^2 \text{ or } \frac{1}{n-1}\left[\sum_{i=1}^{n} x_i^2 - n\overline{X}^2\right] \text{ or } \frac{1}{n-1}\left[\sum x_i^2 - \frac{\left(\sum x_i\right)^2}{n}\right]$$

The **standard deviation** of a data set is defined as the positive square root of the variance of the data (regardless of whether the data are for the population or for the sample). The **sample coefficient of variation** is defined as the standard deviation per unit mean (in percentages):

$$\text{CV} = \frac{s}{\overline{X}} \times 100\%$$

The ***pth* percentile** of a set of data is defined as the value for which (at most) *p*% of the data are less than that value and at most (100-*p*)% of the data are greater than that value. The 25th, 50th and the 75th percentiles are called the **quartiles**.

The skewness (*sk*) is a numerical measure that assesses the lack of symmetry of the distribution. If *sk* is positive, the distribution is positively (right) skewed and if *sk* is negative, the distribution is negatively (left) skewed. If *sk* is zero, then the distribution will be symmetric. Kurtosis (*K*) is another numerical measure that can be used to assess the relative peakedness or flatness of the distribution compared to the normal distribution. If K=0, the distribution is normal. If *K* is positive, the distribution is relatively peaked and if *K* is negative, the distribution is relatively flat.

4.3 Approximate mean and variance of grouped data

In this section, we look at the measurement of the two most important descriptive measures, mean and variance (or standard deviation), when the data are already grouped into a frequency distribution. The approximate mean for grouped data, **group mean,** is defined as,

$$\text{Group mean } \overline{X}_G = \frac{\sum_{i=1}^{K} f_i M_i}{n}$$

and the approximate variance for grouped data, **group variance**, is defined as,

$$\text{Group variance } s_G^2 = \frac{1}{n-1}\left[\sum_{i=1}^{K} f_i M_i^2 - n\overline{X}_G^2\right]$$

where M_i is the mid-point of the ith class, i=1, 2, ..., *K*, f_i is the frequency of that class and *K* is the number of classes.

The **group standard deviation** is approximated by the square root of the group variance.

Aim: At the completion of the examples in this chapter you should be able to use **EXCEL** to calculate the numerical descriptive summary measures such as mean, median, mode, variance, standard deviation, maximum, minimum, range, quartiles, percentiles, percentage rank of an observation, skewness, kurtosis, and approximate mean, variance and standard deviation of a grouped data set.

The following Example 4.1 explains the various methods of calculating the summary statistics using EXCEL.

4.4 Example 4.1: Calculating descriptive numerical summary measures

(Ref: Exercise 3.18, p.56, Australian Business Statistics, Selvanathan et al., 1994)

The **sample** results (X) for 30 students from a statistics exam are as follows:

75	79	58	73	82	94
61	77	54	77	65	67
62	61	64	45	58	86
66	83	70	91	48	78
86	66	52	80	59	55

use EXCEL to perform the following tasks.

Task 1 Type the title 'Exam Results' in cell A1 and the data in cells A2 to A31, and save the data in a file with name 'A:EX41.XLS'.

Task 2 Calculate $\sum x_i$, $\sum x_i^2$ and $\sum(x_i - \overline{X})^2$.

Task 3 Using the values obtained in Task 2, calculate the following:

$$\text{Sample mean } \overline{X} = \frac{\sum x_i}{n}$$

$$\text{Var1} = \frac{1}{n-1}\sum_{i=1}^{n}(x_i - \overline{X})^2$$

$$\text{Var2} = \frac{1}{n-1}\left[\sum_{i=1}^{n} x_i^2 - n\overline{X}^2\right]$$

$$\text{Var3} = \frac{1}{n-1}\left[\sum_{i=1}^{n} x_i^2 - \frac{\left(\sum x_i\right)^2}{n}\right]$$

Comment on the relationship between Var1, Var2 and Var3. Calculate the sample standard deviation.

Task 4 Calculate the mean, median, mode, variance, standard deviation, minimum, maximum, range, the three quartiles, interquartile range, 20th and 90th percentiles, skewness and kurtosis for the sample data.

Task 5 Determine the percentage rank of the student whose score is 80.

Task 6 Set up a frequency distribution for the exam scores.

Task 7 Using the frequency distribution, calculate the (approximate mean and variance)

$$\text{Group mean} \quad \overline{X}_G = \frac{\sum_{i=1}^{K} f_i M_i}{n}$$

$$\text{Group variance} \quad s_G^2 = \frac{1}{n-1}\left[\sum_{i=1}^{K} f_i M_i^2 - n\overline{X}_G^2\right]$$

and the group standard deviation, where M_i is the mid-point of the ith class and f_i is the frequency of the ith class. Compare the group mean and the group standard deviation with the means and standard deviations obtained in Task 3 and Task 4.

Task 8 Save the EXCEL workbook on your disk as 'A:EX41.XLS' and print the file.

Solution to Example 4.1

Task 1 Open a new EXCEL worksheet and type the column heading 'Exam Results' in cell A1 and adjust the column width to fit the heading. Type data in cells A2 to A31. Save the file on your disk in drive A as **A:EX41.XLS**.

Task 2 Let us denote the exam results by X. To calculate the required sums of the data set, you can use the icon **Paste Function** on the *Standard toolbar*.

Calculating the sum of X: Σx_i

To calculate the *SUM* of the given 30 exam results, follow Steps 1-5:

Step 1 In cell A32, type the label 'Sum of X =' to identify the sum and adjust the column width to fit the label. (Note that the selection of cell A32 and typing the text is optional, you can select any one of the cells to put the answer for the sum.)

Step 2 Click on cell B32 and move the cursor to the *Standard toolbar* and click on the icon **Paste Function**. A dialogue box will appear on the screen (see Figure 4.1).

Step 3 From the dialogue box, select the *Function Category* option **All**. A list of mathematical and statistical functions will appear on the right-hand side.

Step 4 Now, to find the *SUM*, choose the *Function Name* option **SUM** and click **OK**. You will see another dialogue box on the screen (see Figure 4.4).

Figure 4.4 SUM Function dialogue box

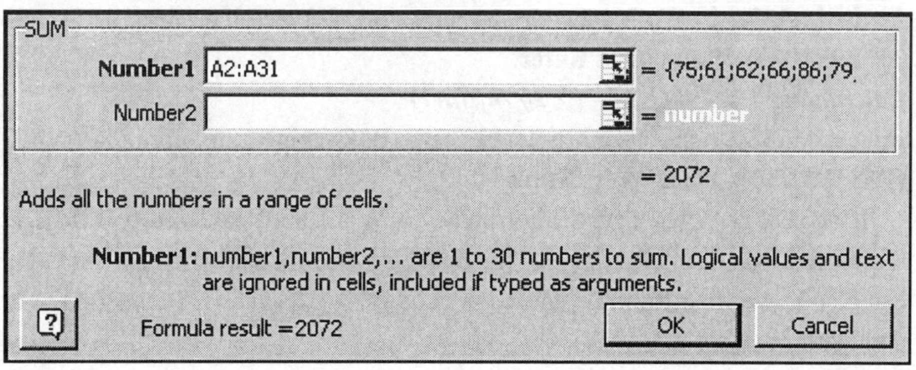

Step 5 On the dialogue box, you will be asked to list the arrays of data to be used to calculate the sum. Type the location of the data array

number1. In this example, the data *array (Number 1)* is A2 to A31. Type **A2:A31** and then click **OK** on the dialogue box.

Short-cut method: Instead of going through Steps 2-5, you can click the mouse on cell B32 and type **=SUM(A2:A31)**. [*Alternatively*, highlight A2:A31 and click the mouse on the icon [Σ] **Autosum** on the *Standard toolbar*. You will see the sum value appearing in cell A32. Now use the pointer to move the value in cell A32 to cell B32 and type the label 'Sum of X' in cell A32.]

Calculating the sum of X^2: $\sum x_i^2$

Type 'Sum X squares=' in cell A33. Then follow the same Steps 2-5 as above, except selecting cell B33 in Step 2 and selecting *Function Name* **SUMSQ** instead of 'SUM' in Step 4.

Short-cut method: Instead of going through Steps 2-5, click the mouse on cell B33 and then type **=SUMSQ(A2:A31)**.

Calculating the sum of deviation squares: $\sum (x_i - \overline{X})^2$

Type 'Deviation squares=' in cell A34. Then follow the same Steps 2-5 as above, except selecting cell B34 in Step 2 and selecting *Function Name* **DEVSQ** in Step 4.

Short-cut method: Instead of going through Steps 2-5, click the mouse on cell B34 and then type **=DEVSQ(A2:A31)**.

Task 3 ***Calculating the mean of X: $\overline{X} = \sum x_i/n$***

In cell A35, type 'Mean =' and click on cell B35. In cell B35, type the formula **=B32/30** and press **Enter**.

Calculating the sample variance[1] of X:

Calculating Var1 = $\sum (x_i - \overline{X})^2/(n-1)$

In cell A36, type 'Var1 ='. In cell B36, type the formula **=B34/29** and press **Enter**.

Calculating Var2 = $(\sum x_i^2 - n\overline{X}^2)/(n-1)$

In cell A37, type 'Var2 ='. In cell B37, type the formula **=(B33-30*B35*B35)/29** and press **Enter**.

Calculating Var3 = $[\sum x_i^2 - (\sum x_i)^2/n]/(n-1)$

In cell A38, type 'Var3 ='. In cell B38, type the formula **=(B33-(B32*B32/30))/29** and press **Enter**.

It should be noted that the formulae for Var2 and Var3 are two different representations of the formula for the ***Sample variance s^2*** defined by Var1. This can also be verified from the entries in cells B36, B37 and B38 that Var1, Var2 and Var3 are the same.

---

[1] To calculate the population variance and standard deviation, the divisor for Var1, Var2 and Var3 should be replaced by N rather than *n-1*. The population standard deviation is calculated as the square root of the population variance.

Calculating the sample standard deviation[1] of X: s =$\sqrt{Variance}$

In cell A39, type 'Standard deviation='. In cell B39, type the formula **=SQRT(B36)** or **=SQRT(B37)** or **=SQRT(B38)** and press **Enter**.

For alternative methods of calculating the mean, variance and standard deviation using the *Paste Function*, see Task 4.

Task 4 Method 1 To calculate most of the summary statistics of a given data set, you can use the icon  **Paste Function** on the *Standard toolbar*. First, in cell C1, type the title 'Summary Measures' and adjust the column width to fit the title.

Calculating the mean, median, mode, variance, standard deviation, maximum, minimum, skewness and kurtosis

To obtain the *MEAN*, follow Steps 1 to 5 as in Task 2 except in Step 1 use cell C2 to type the text 'Mean =', in Step 2 use cell D2 to store the mean value and in Step 4 choose the *Function Name* option **AVERAGE**. *Alternatively*, instead of going through Steps 2-5, you can simply click the mouse on cell D2 and type **=AVERAGE(A2:A31)**.

To obtain the *MEDIAN*, follow Steps 1 to 5 of Task 2 as before except in Step 1 use cell C3 to type the text 'Median =', in Step 2 use cell D3 to store the median value and in Step 4 choose the *Function Name* option **MEDIAN**. *Alternatively*, after Step 1, click the mouse on cell D3 and type **=MEDIAN(A2:A31)**.

Repeat Steps 1-5 of Task 2 to calculate the *MODE*, *VARIANCE*, *STANDARD DEVIATION, MAXIMUM, MINIMUM, SKEWNESS* and *KURTOSIS* by selecting different cells in Step 1 (cells C4, C5, C6, C7, C8, C9 and C10 instead of C2) to type the labels; use different cells in Step 2 (cells D4, D5, D6, D7, D8, D9 and D10 instead of D2) to store the values; in Step 3 select the *Function Category* option **All** or **Statistical;** and in Step 4 choose the *Function Name* options **MODE, VAR,**[2] **STDEV,**[2] **MAX, MIN, SKEW** and **KURT**; respectively. *Alternatively*, after Step 1, type **=MODE(A2:A31)** in cell D4; type **=VAR(A2:A31)** in cell D5, type **=STDEV(A2:A31)** in cell D6; type **=MAX(A2:A31)** in cell D7; type **=MIN(A2:A31)** in cell D8; type**=SKEW(A2:A31)** in cell D9; type **=KURT(A2:A31)** in cell D10.

Calculating the range

To obtain the *RANGE* (which is the difference between the maximum and minimum values of a data set), in EXCEL there is no

---

2 If you are using all the observations from the population (instead of a sample), to calculate the variance and standard deviation, the corresponding functional names to be selected are **STDEVP** and **VARP**, respectively.

straight forward option available in the dialogue box as for *MEAN, MEDIAN* etc. You have to write your own mathematical formula using the available standard mathematical functions MAXIMUM and MINIMUM in **EXCEL**. First type the label 'Range =' in a cell (for example, in C11). In cell D11, type the formula **=D7-D8** or **=MAX(A2:A31)-MIN(A2:A31)**. Now you will see the actual RANGE value appearing in cell D11.

Calculating the three quartiles

To obtain the *1st Quartile* which is also the *25th Percentile*, type 'First Q=' in cell C12, click on cell D12 and click on the icon *fx* **Paste Function** and select option **All** or **Statistical** as in Steps 2 and 3 above. In Step 4, select the *Function Name* option **QUARTILE** and click **OK**. The resulting dialogue box will request two entries (see Figure 4.5). For the first entry '*Array*', type the data range **A2:A31** and, for the second entry '*Quart*', type **1**. Then click **OK** on the dialogue box. *Alternatively,* in cell D12 type **=QUARTILE(A2:A31,1)**. Similarly, we can also obtain the *2nd Quartile* (which is also the *50th Percentile*) and the *3rd Quartile* (which is also the *75th Percentile*) as **=QUARTILE(A2:A31,2)** and **=QUARTILE (A2:A31,3)**, and store it in cells D13 and D14, respectively with the corresponding labels, 'Second Q=' and 'Third Q=' in cells C13 and C14, respectively.

Figure 4.5 QUARTILE Function dialogue box

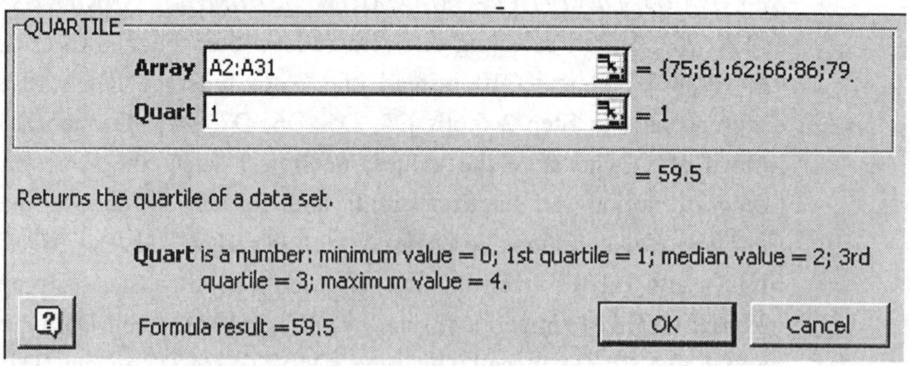

Calculating the interquartile range

To obtain the interquartile range (which is the difference between the third and the first quartiles) there is no straight forward option available. In cell C15, type the label 'Interquartile range=' and in cell D15, type **=D14-D12**.

Calculating the percentiles

In cell C16, type the label '20th Percentile=' and click on cell D16. To calculate the Percentile values, follow the same steps as for the Quartiles, except that in Step 4, select the *Function Name* option **PERCENTILE** (instead of QUARTILE). In the resulting dialogue box, for the first entry '*Array*', type the data range **A2:A31**, and for the second entry '*k*', type, for example, **.20** for the 20th percentile. The 20th percentile value will appear in cell D16. *Alternatively,* in cell D16, type **=PERCENTILE(A2:A31,.2)**. Similarly, we can obtain the 90th percentile. Store the label in cell C17 as '90th Percentile=' and the corresponding value in cell D17 as **=PERCENTILE(A2:A31,.9)**.

Figure 4.6 presents the output in cells Cl to D17 (we shall discuss the entries in cells C18 to D18 in Task 5).

Figure 4.6 Descriptive summary measures

	C	D
1	Summary measures	
2	Mean=	69.07
3	Median=	66.50
4	Mode=	61.00
5	Var=	167.03
6	SD=	12.92
7	Max=	94.00
8	Min=	45.00
9	Skewness=	0.09
10	Kurtosis=	-0.83
11	Range=	49.00
12	First Q=	59.50
13	Second Q=	66.50
14	Third Q=	78.75
15	20th percentile=	58.00
16	90th percentile=	86.00
17	Percentagr rank of 80 =	0.793

Method 2 Most of the numerical measures such as the MEAN, MEDIAN, MODE, VARIANCE, STANDARD DEVIATION, MAXIMUM, MINIMUM, RANGE, SKEWNESS and KURTOSIS can also be calculated using **Data Analysis...** under **Tools** on the *Menubar* (see Figure 3.1).

In this method, first click on **Tools** on the *Menubar* and then select the **Data Analysis...** option.[3] A list of **Analysis Tools** will be

3 If **Data Analysis** is not on the list, select the **Add-In** option, click on the square against the *Analysis Tookpak* and click **OK**. Now if you click on **Tools** in the *Menubar*, the menu will have **Data Analysis...** included as its last option.

displayed (see Figure 4.2). Select the **Descriptive Statistics** option and click **OK**. A dialogue box will appear (see Figure 4.3). For the *Input range*, type **A2:A31**; click on the circle for the *Output range,* click on the adjacent space and enter the starting cell location for the output as **A45**; click on the square for **Summary Statistics**; and click **OK** (see Figure 4.7). Starting from cell A45, the output will display a number of summary statistics (see Figure 4.8). Note that this option automatically assumes that the data are for the sample and gives the sample standard deviation and the sample variance.

Figure 4.7 Descriptive Statistics dialogue box

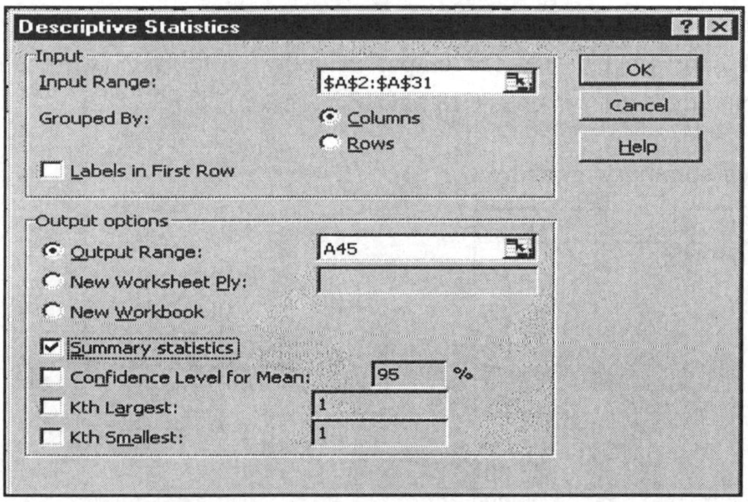

Figure 4.8 Descriptive Statistics results

Column 1	
Mean	69.06666667
Standard Error	2.359589548
Median	66.5
Mode	61
Standard Deviation	12.92400422
Sample Variance	167.0298851
Kurtosis	-0.831391555
Skewness	0.086245626
Range	49
Minimum	45
Maximum	94
Sum	2072
Count	30

Task 5 ***Calculating the percent rank***

To determine the percentage rank of mark 80, we use the **Paste Function**. In cell C18, type the label 'Percent Rank of score 80=' and click on cell D18. Then, click on the icon *fx* **Paste Function** and select the *Function Category* **All** or **Statistical**. Then select the *Function Name* '**PERCENTRANK**' and click **OK**. Now you will see the dialogue box for PERCENTRANK (see Figure 4.9).

Figure 4.9 PERCENTRANK Function dialogue box

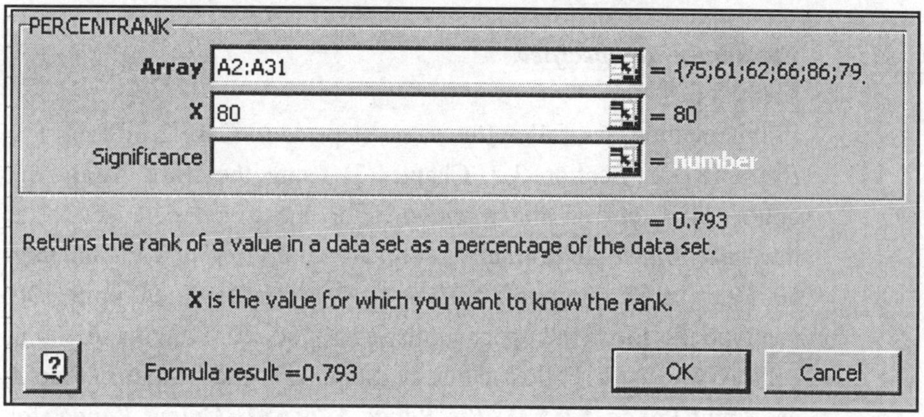

In the dialogue box, you will be asked to enter the location of the data array and a value for *x* for which we want to know the rank. For this task, we enter **A2:A31** for '*Array*' and **80** for 'X'. Now click **OK**. The resulting value will appear in cell D18 (see Figure 4.6). *Alternatively,* in cell D18, type the formula **=PERCENTRANK(A2:A31,80)**.

Task 6 ***Constructing a frequency distribution for the exam scores***

Case 1 ***Class intervals not specified***

In this case, we allow EXCEL to automatically determine the class intervals. We construct the frequency distribution by using the same steps as in Case 1 of Task 1 in Example 3.1, (Section 3.2, Chapter 3) using the *Histogram* under the **Data Analysis...** option in **Tools** on the *Menubar*. To construct a frequency distribution of the exam scores, follow the same Steps 1-3 of Section 3.2 (page 28), except that in Step 2, use *Input Range* **A2:A31**; *Output Range* (for example) **A65**; and **do not** tick *Chart Output*. [Note that ticking *Chart Output* will result in the Histogram as well.] This will result in the frequency distribution with automatic class interval specification (see Figure 4.10).

Figure 4.10 Automatically specified class intervals and frequencies

Bin	Frequency
45	1
54.8	3
64.6	8
74.4	6
84.2	8
More	4

Case 2 ***Class intervals specified***

Method 1

In this method, we follow the same steps as in Case 2 of Task 1 in Example 3.1 (Section 3.2, Chapter 3) using the **Data Analysis...** option in **Tools** on the *Menubar*. For the given data set, class intervals you would normally set up are 40 up to *(but not including)* 50, 50 up to 60, 60 up to 70, 70 up to 80, 80 up to 90, 90 up to 100. We type the **Bins** (the upper limits) as 50, 60, 70, 80, 90 and 100 in cells A76 to A81. Follow the same Steps 1-3, except that in Step 2, usc *Input Range* **A2:A31**; *Bin Range* **A76:A81**; *Output Range* (for example) B75; and **do not** tick *Chart Output*. [Note that ticking *Chart Output* will result in the Histogram as well.] This will result in the frequency distribution with the specified Bin values (see Figure 4.11).

Figure 4.11 Specified class intervals and frequencies

Bin	Frequency
50	2
60	6
70	9
80	7
90	4
100	2
More	0

Method 2

In this method, we use the Paste Function to set up the frequency distribution. To compute the frequency in each class interval, we need to set up the class intervals in a special way. For the given data set, class intervals we would normally set up are 40 up to (but not including) 50, 50 up to 60, 60 up to 70, 70 up to 80, 80 up to 90, 90 up to 100. However, to use EXCEL, in a column of the worksheet (for example, in cells C20-C25), type the upper limits of the class intervals only and type column heading 'Bin Array' in cell C19, i.e.,

Bin array	Frequency f
50	
60	
70	
80	
90	
100	

We will nominate cells E20 to E25 to store the frequencies. Type column heading 'Frequency f' in cell E19. Next click the mouse on cell E20. Then click the mouse on the icon **Paste Function** on the *Standard toolbar*. In the dialogue box that appears on the screen select *Function Category* **Statistical** or **All** and from the list of functions appearing on the right hand side, select *Function Name* **FREQUENCY**. Then click **OK**. You will see another dialogue box which will ask for two sets of entries (see Figure 4.12), one is the *Data array* and the other is the *Bins array*. For the blank space in the *Data_array*, type **A2:A31**. Click the mouse on the blank space in the *Bins_array* (locations of class end points) and type **C20:C25**. After typing these two entries, click **OK**. Now you will see the frequency for the first class interval appearing in cell E20. You will also notice a formula in the *Formula bar* which reads as **=FREQUENCY(A2:A31,C20:C25)**. [***Short cut method:*** Instead of going through the various steps in **Paste Function**, in cell E20, just type **=FREQUENCY(A2:A31,C20:C25)**.]

Figure 4.12 FREQUENCY Function dialogue box

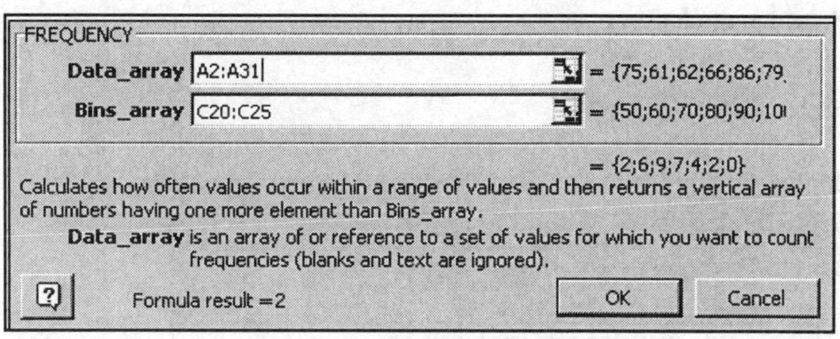

To obtain the frequencies for the remaining classes, first highlight the frequency cells E20 to E25. You will be able to see the formula being displayed on the *Formula bar.* Now click the mouse at the end of the formula [i.e. just after C25)]. Then simultaneously press keys **CTRL+SHIFT+ENTER** on the keyboard. Now you will see the frequencies for cells E21 to E25 appearing in these cells. See cells C19 to E26 of Figure 4.13 for the output. Save the file **A:EX41.XLS**.

Task 7 ***Calculating the approximate mean, variance and standard deviation***

Note from Task 6 that cells E20 to E25 contains the frequencies (*f*). Let ***M*** denotes the midpoints of the class intervals.

Calculating the sums $\Sigma f_i M_i$ ***and*** $\Sigma f_i M_i^2$

In cell F19, G19 and H19, type the column headings 'Midpoint M', 'fxM', and 'fxMxM'. In cells F20 to F25, type the mid-points, **45, 55, 65, 75, 85, 95**. In cell G20, type the formula **=E20*F20** and copy cell G20 onto cells G21 to G25. In cell H20, type the formula **=G20*F20** and copy cell H20 onto cells H21 to H25. To sum the frequencies in E20 to E25, highlight E20 to E25 and click on the icon Σ **Autosum** on the *Standard toolbar* and press **Enter**. The sum, which should be the total number of observations (***n***) will appear in cell E26. To sum the cell entries G20 to G25, highlight G20 to G25 and click on the icon Σ **Autosum** on the *Standard toolbar* and press **Enter**. The sum will appear in cell G26. Similarly, sum the entries in cells H20 to H25, which will appear in cell H26.

Calculating the group mean: $\bar{X}_G = \Sigma f_i M_i / n$

In cell E27, type the label 'Group mean=', and in cell F27, type the formula **=G26/E26** and press **Enter**.

Calculating the group variance: $s_G^2 = \left(\sum f_i M_i^2 - n\bar{X}_G^2\right)/(n-1)$

In cell E28, type the heading 'Group variance=', and in cell F28, type the formula **=(H26-E26*F27*F27)/(E26-1)** and press **Enter**.

Calculating the group standard deviation: $s_G = \sqrt{Group\ var\ iance}$

In cell E29, type the label 'Group SD=', and in cell F29, type the formula **=SQRT(F28)** and press **Enter**.

Figure 4.13 shows the final output.

Figure 4.13 Group Mean, Group Variance and Group Standard Deviation

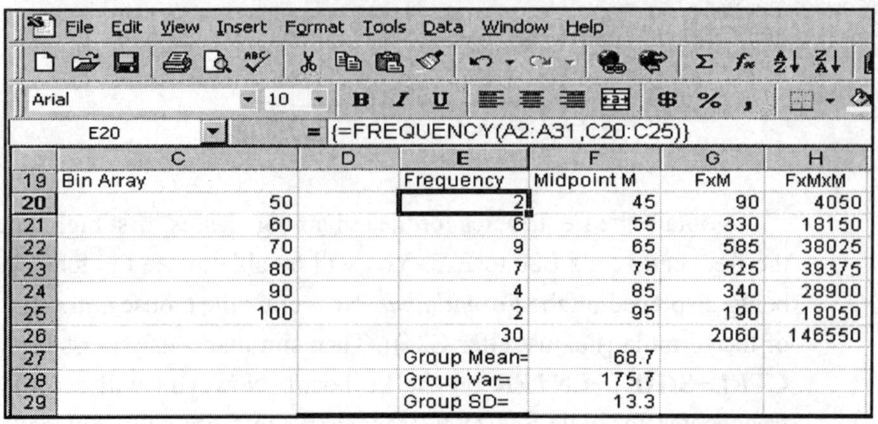

	C	D	E	F	G	H
19	Bin Array		Frequency	Midpoint M	FxM	FxMxM
20	50		2	45	90	4050
21	60		6	55	330	18150
22	70		9	65	585	38025
23	80		7	75	525	39375
24	90		4	85	340	28900
25	100		2	95	190	18050
26			30		2060	146550
27			Group Mean=	68.7		
28			Group Var=	175.7		
29			Group SD=	13.3		

Task 8 Save the file again on your disk in drive A as **A:EX41.XLS**. Print the file using the **Print** option in the **File** menu on the *Menubar*.

4.5 Exercises to Chapter 4

(The following exercises are based on exercises from Chapter 3 of Selvanathan et al., Australian Business Statistics, 1994)

Exercise 4.1

The dividend yields for 20 Australian industries are expressed as percentages in the following table.

3.3	3.9	3.9	5.3
2.2	4.8	4.4	3.0
4.9	2.1	3.1	7.0
2.2	3.9	3.8	3.9
2.9	3.0	0.6	10.4

Use EXCEL to perform the following tasks.

Task 1 Type the title 'Dividend Yields' in cell A1 and the data in cells A2 to A21.

Task 2 Calculate $\sum x_i$, $\sum x_i^2$ and $\sum (x_i - \overline{X})^2$.

Task 3 Using the values calculated in Task 2, calculate the following:

$$\text{Sample mean } \overline{X} = \frac{\sum x_i}{n}$$

$$\text{Var1} = \frac{1}{n-1}\sum_{i=1}^{n}(x_i - \overline{X})^2$$

$$\text{Var2} = \frac{1}{n-1}\left[\sum_{i=1}^{n} x_i^2 - n\overline{X}^2\right]$$

$$\text{Var3} = \frac{1}{n-1}\left[\sum x_i^2 - \frac{\left(\sum x_i\right)^2}{n}\right]$$

Comment on the relationship between Var1, Var2 and Var3. Calculate the standard deviation.

Task 4 Calculate the mean, median, mode, variance, standard deviation, minimum, maximum, range, the three quartiles, interquartile range, 30th and 80th percentiles, skewness and kurtosis.

Task 5 Determine the percentage rank of the industry whose dividend yield is 5.3.

Task 6 Set up a frequency distribution for the dividend yields.

Task 7 Using the frequency distribution calculate the (approximate mean and variance)

Group mean $$\overline{X}_G = \frac{\sum_{i=1}^{K} f_i M_i}{n}$$

Group variance $$s_G^2 = \frac{1}{n-1}\left[\sum_{i=1}^{K} f_i M_i^2 - n\overline{X}_G^2\right]$$

and the group standard deviation, where M_i is the mid-pint of class i and f_i is the frequency of class i. Compare the group mean and the group standard deviation with the means and standard deviations obtained in Task 3 and Task 4.

Task 8 Graph the frequency histogram for the dividend yields.

Task 9 Save the EXCEL workbook on your disk as 'A:DIVID.XLS' and print the file.

Exercise 4.2

A large investment firm in Sydney wants to review the distribution of the ages of its stockbrokers. The firm feels that this information will be useful in developing plans relating to recruitment and retirement options. The ages of a sample of 25 brokers were recorded as follows:

50	64	32	55	41
44	24	46	58	47
36	52	54	44	66
47	59	51	61	57
49	28	42	38	45

Use EXCEL to perform the following tasks.

Task 1 Find the median age.

Task 2 Find the lower quartile and the upper quartile of the ages.

Task 3 Find the 80th percentile of the ages.

Task 4 Does the firm have any reason to be concerned about the distribution of ages of its brokers?

Task 5 Compute the mean, variance and standard deviation of the sample of data.

Task 6 Compute the range of the data. Compute the range approximation to the standard deviation of the data.

Task 7 Construct a frequency distribution for the data using five class intervals and the value 20 as the lower limit of the first class.

Task 8 Approximate the mean and the variance of ages based on the frequency distribution constructed in Task 7.

Task 9 Save your workbook with filename 'A:AGE.XLS' on your disk and print the file.

Chapter 5

Probability

In situations which involves uncertainty, probability is used as a tool by most decision-makers to make rational judgements. *Probability* is used to measure the likelihood of occurrence of an event. There are three general approaches to assigning probabilities to events: classical approach, relative frequency approach, and subjective approach. The *classical approach,* also called the equally likely approach, is based on the symmetrical nature of the experiment; the *relative frequency approach* is based on the long-run relative frequency of occurrence of the experiment; and the *subjective approach* reflects the degree to which we believe that the outcome will occur. Among the three approaches, EXCEL can be used to calculate probabilities based on the relative frequency approach. Using the relative frequency approach, the probability of an outcome is defined as

$$\begin{aligned} \text{P(outcome)} &= \frac{\text{Number of times a particular outcome occurred}}{\text{Total number of trials}} \\ &= \frac{x}{n} \end{aligned}$$

where x is the frequency of occurrence of the outcome in n repeated identical trials.

5.1 Joint, marginal and conditional probabilities

A *joint probability* gives the probability that an observation will possess two or more characteristics simultaneously. A *marginal probability* gives the proportion of observations that possess a single characteristic.

The probability that some event occurs given that some other event has already occurred is called the conditional probability of that event. The *conditional probability* of event A given that event B has occurred is defined as

$$P(A|B)=\frac{P(A\cap B)}{P(B)}$$

It should be noted that the term in the denominator is the marginal probability of B and the term in the numerator is the joint probability of A and B. Therefore, the conditional probabilities can be calculated by using the joint and marginal probabilities.

EXCEL can be used to set-up the joint frequency table and joint probability table (in percentages). To obtain these tables for a given raw data set with two or more characteristics, we first select **Data** on the *Menubar* (see Figure 5.1) and then select option **Pivot Table Report…** which will give a Pivot Table Wizard dialogue box with 4 steps. At the completion of the 4 steps, depending on the options selected, you will obtain either a joint frequency table or a joint probability table (in percentage). The last column and the last row with the title 'Grand Total' of the joint frequency table represent the marginal frequencies and those of the joint probability table represent the marginal probabilities (in percentages). The conditional probabilities can be calculated by using the relevant entries from the joint and marginal probabilities from the joint-probability table.

Figure 5.1 Menubar and Data menu

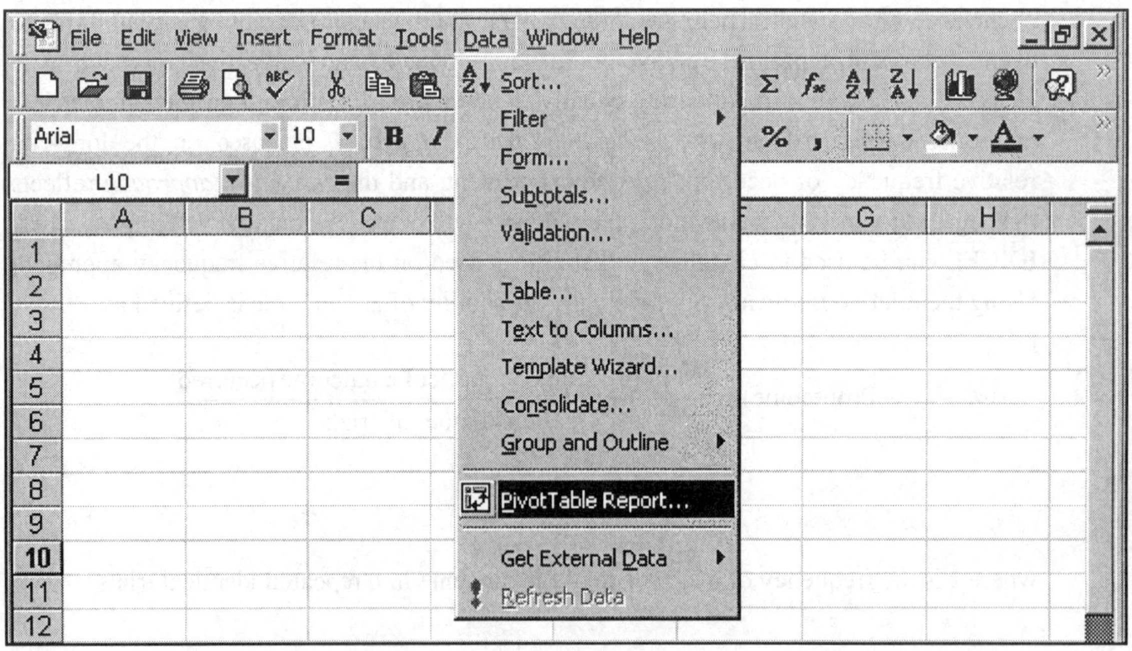

Aim: At the completion of the examples in this chapter you should be able to use **EXCEL** to set-up joint and marginal probability tables for events based on raw data. Based on these probabilities, you will also be able to calculate probabilities for various other (unconditional and conditional) events.

5.2 Example 5.1: Constructing joint and marginal probability tables, and calculating the joint, marginal and conditional probabilities

The following are the survey data of 30 people interviewed at a Shopping Mall regarding their smoking habit (smoker/non-smoker) and their health condition (have cancer/have heart problem/good health), together with their ID numbers.

ID	HABIT	HEALTH CONDITION
1	Smoker	Cancer
2	Non-Smoker	Heart Problem
3	Non-Smoker	Good Health
4	Non-Smoker	Good Health
5	Smoker	Good Health
6	Non-Smoker	Cancer
7	Non-Smoker	Good Health
8	Smoker	Cancer
9	Smoker	Cancer
10	Smoker	Heart Problem
11	Non-Smoker	Cancer
12	Non-Smoker	Good Health
13	Non-Smoker	Good Health
14	Smoker	Cancer
15	Non-Smoker	Heart Problem
16	Non-Smoker	Good Health
17	Smoker	Good Health
18	Smoker	Cancer
19	Smoker	Cancer
20	Non-Smoker	Good Health
21	Non-Smoker	Good Health
22	Non-Smoker	Cancer
23	Smoker	Cancer
24	Non-Smoker	Good Health
25	Non-Smoker	Good Health
26	Smoker	Good Health
27	Smoker	Cancer
28	Non-Smoker	Good Health
29	Non-Smoker	Good Health
30	Non-Smoker	Good Health

Use EXCEL to perform the following tasks.

Task 1 Construct two separate marginal frequency tables in absolute frequency form for HABIT and HEALTH CONDITION.

Task 2 Construct two separate marginal relative frequency tables in percentage form for HABIT and HEALTH CONDITION.

Task 3 Draw bar charts to show the above frequencies (from Tasks 1 and 2) in absolute and percentage form.

Task 4 Set up a two-variable joint frequency table (in absolute form) and joint probability table (in relative frequency form) for HABIT and HEALTH CONDITION.

Task 5 Draw a bar chart for the joint relative frequencies.

Task 6 Use the *Joint Probability Table* from Task 4 to locate/derive the answers to the following questions. If one person surveyed is selected at random,

a. What is the probability that the person selected is a *non-smoker* and has *heart problem*?

b. What is the probability that the person selected is a *smoker* and has *heart problem*?

c. What is the probability that the person selected is a *smoker* and has *cancer*?

d. What is the probability that the person selected is a *smoker*?

e. What is the probability that the person selected is in *good health*?

f. What is the probability that the person selected has *cancer*?

g. What is the probability that the person selected has *heart problem or cancer* and is a *smoker*?

h. What is the probability that the person selected has *heart problem or cancer* and is a *non-smoker*?

Task 7 a. Given that the person selected has *cancer*, what is the probability that he/she is *a smoker*?

b. If the person selected is *a non-smoker*, what is the probability that he/she has *heart problem*?

c. If the person selected has *cancer or heart problem*, what is the probability that he/she is a *smoker*?

Task 8 Save the EXCEL workbook as 'A:EX51.XLS' on your disk and print the file.

Solution to Example 5.1

Task 1 Open a new EXCEL worksheet and type the column titles and the data in cells A1 to C31.

Constructing the marginal frequency table: HABIT

Click the mouse in any cell in the data area. Then select **Data** on the *Menubar* and select option **Pivot Table Report....** Now you will see a **PivotTable Wizard - Step 1 of 4** dialogue box appearing on the screen (see Figure 5.2).

Figure 5.2 PivotTable dialogue box - step 1

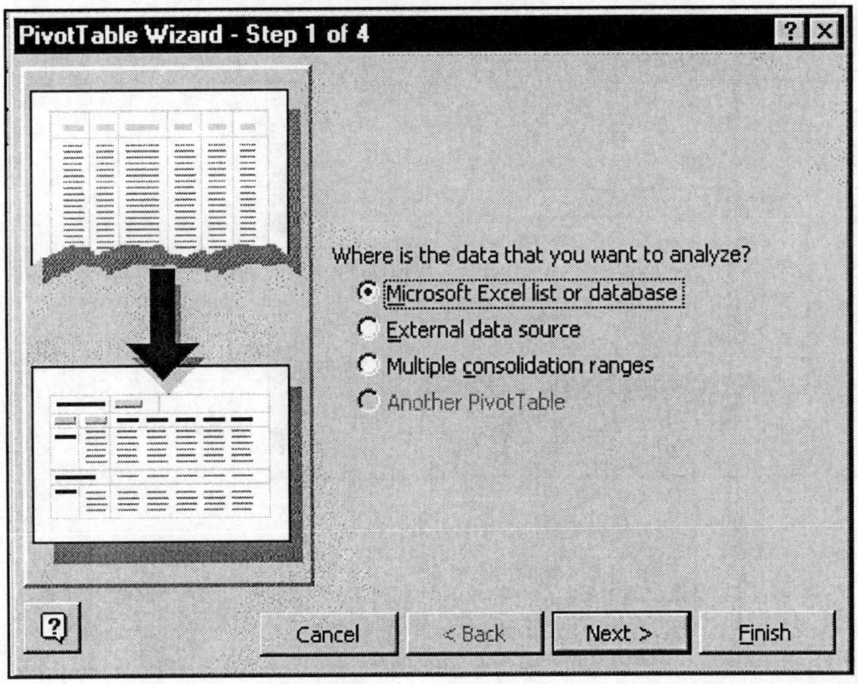

In Step 1, click on the circle for **Microsoft Excel List or Database** and then click **Next>**. Now **PivotTable Wizard - Step 2 of 4** dialogue box will appear (see Figure 5.3).

Figure 5.3 PivotTable dialogue box - step 2

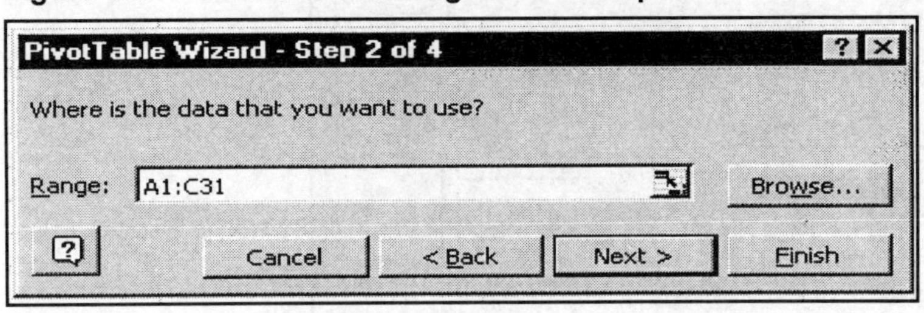

In Step 2, the *Range* will show the data range as $A$1:$C$31. If the data range is not displayed, type the data range. Click **Next>** to go to Step 3. In Step 3 you will see a dialogue box as shown in Figure 5.4.

As you can see from Figure 5.4, the three titles ID, HABIT, HEALTH CONDITION in cells A1, B1 and C1, respectively, are displayed as three buttons on the right. In this step, click on the **HABIT** button and drag it to the **ROW** area on the left and release. Next, click on the **ID** button and drag it to the **DATA** area in the center and release. The new label should be displayed as

Figure 5.4 PivotTable dialogue box - step 3

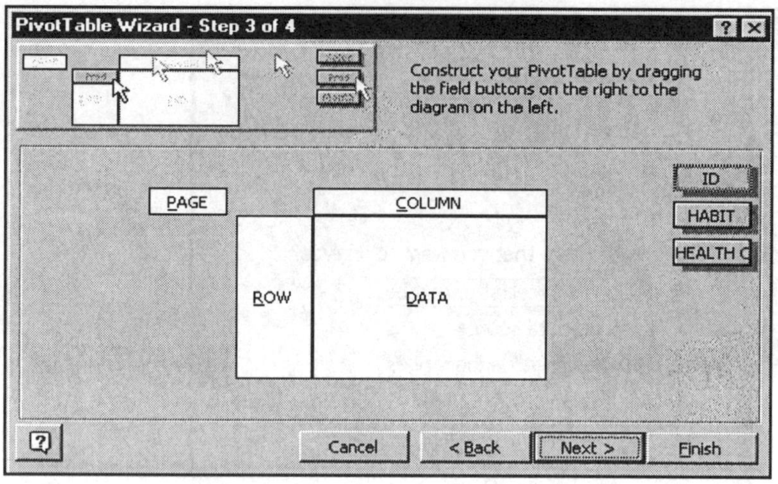

Count of ID. If not, double click on the label **Sum of ID** and a **PivotTable Field** dialogue box will appear. In the *Summarise by:* list, select **Count** (see Figure 5.5) and click **OK**.

Figure 5.5 PivotTable Field dialogue box

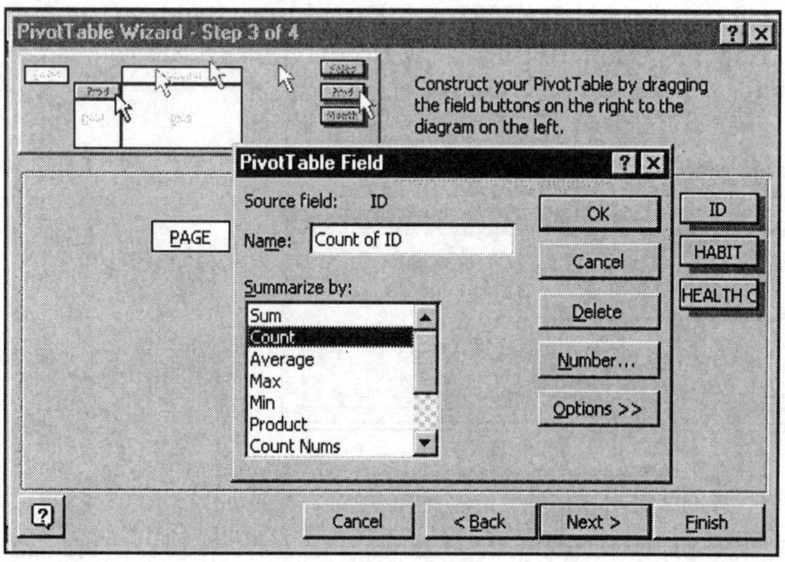

Now click **Next>** to go to Step 4. In Step 4 of the Pivot Table Wizard, click the circle against *Existing Worksheet* and type a cell location to write the output (for example, **A35**), see Figure 5.6. Finally, click **Finish**. The resulting output is displayed in Figure 5.7.

Figure 5.6 PivotTable dialogue box - step 4

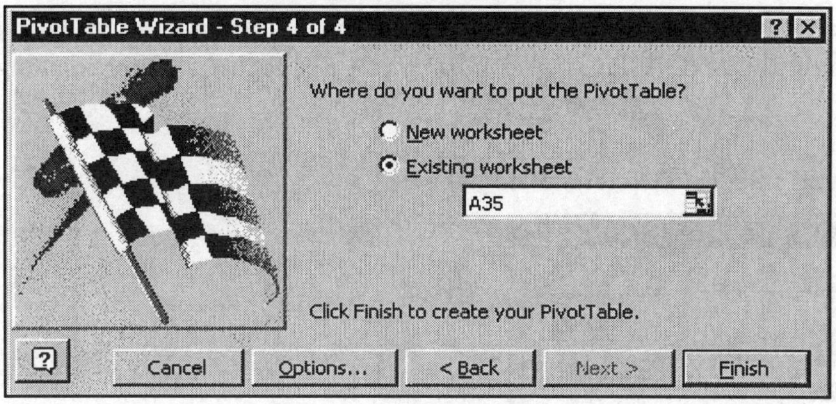

Figure 5.7 Marginal frequency table - HABIT

	A	B
35	Count of ID	
36	HABIT	Total
37	NONSMOKER	18
38	SMOKER	12
39	Grand Total	30

Constructing the marginal frequency table: HEALTH CONDITION

Follow the same steps as for constructing the marginal frequencies for HABIT, except in Step 3, where you need to drag the **HEALTH CONDITION** button instead of HABIT, and in Step 4, select cell **A45** instead of cell A35. After Step 2, you may find a Microsoft Excel dialogue box appearing on the screen. Click **No**. See Figure 5.8 for the final output.

Figure 5.8 Marginal frequency table - HEALTH CONDITION

	A	B
45	Count of ID	
46	HEALTH CONDITION	Total
47	CANCER	11
48	GOOD HEALTH	16
49	HEART PROBLEM	3
50	Grand Total	30

Task 2 ***Constructing the marginal relative frequency table: HABIT***

To convert the frequencies into percentages (relative frequencies × 100) repeat Steps 1-4 as in Task 1 except two minor variations in Steps 3 and 4. In Step 3, after dragging **ID** into the center **DATA** area as before, double click on the new ID label to obtain the **PivotTable Field** dialogue box. First select **Count** from the *Summarise by*: list. Then click on **Options>>**. An extended dialogue box will appear. Click on the **down arrow** on '*Show Data as:*' and select (**% of**

column) and click **OK** (see Figure 5.9). Then drag the **HABIT** button into the **ROW** area. Follow as in Task 1 of Step 3. In Step 4, instead of selecting cell A35, select cell **A55** for *PivotTable Starting cell.* See Figure 5.10 for resulting output.

Figure 5.9 PivotTable Field Options dialogue box

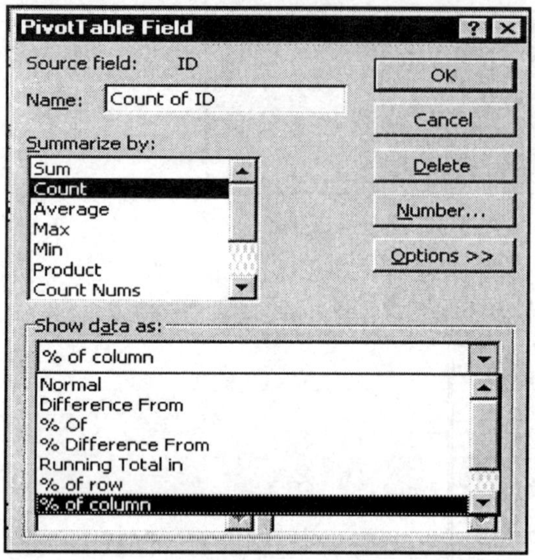

Figure 5.10 Marginal relative frequency table - HABIT

	A	B
55	Count of ID	
56	HABIT	Total
57	NONSMOKER	60.00%
58	SMOKER	40.00%
59	Grand Total	100.00%

Constructing the marginal relative frequency table: HEALTH CONDITION

Repeat the same as above except selecting **HEALTH CONDITION** instead of HABIT in Step 3 and selecting cell **A66** for *PivotTable Starting cell* in Step 4. See Figure 5.11 for the resulting output.

Figure 5.11 Marginal relative frequency table - HEALTH CONDITION

	A	B
66	HEALTH CONDITION	Total
67	CANCER	36.67%
68	GOOD HEALTH	53.33%
69	HEART PROBLEM	10.00%
70	Grand Total	100.00%

Task 3 ***Drawing a bar chart for the marginal frequencies***

To draw a bar chart for the frequencies of HABIT, highlight the frequencies in cells **A37:B38** and use **ChartWizard** on the *Standard Toolbar* and follow Steps 1-4 to draw the graph (for example in Step 1, select *Chart type* **Bar** and *Chart sub-type* **Clustered bar with a 3-D visual effect**, first selection in the second row, and in Step 3 give a chart title). See Figure 5.12 for the resulting bar chart. To draw a bar chart for the frequencies of HEALTH CONDITION, follow the same steps except select data in cells **A47:B49**. The corresponding bar chart is presented in Figure 5.13.

Figure 5.12 Bar chart for marginal frequencies - HABIT

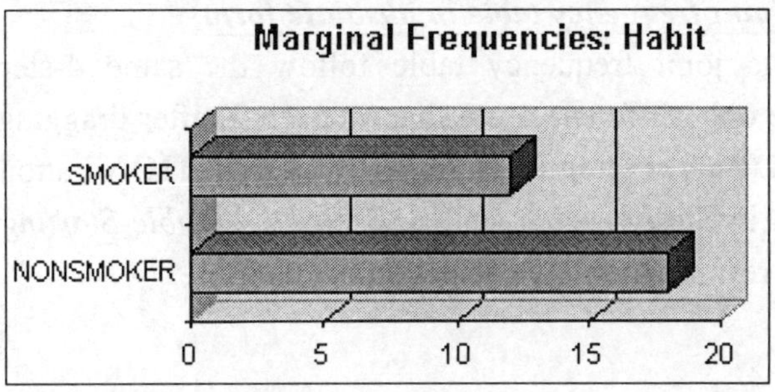

Figure 5.13 Bar chart for marginal frequencies - HEALTH CONDITION

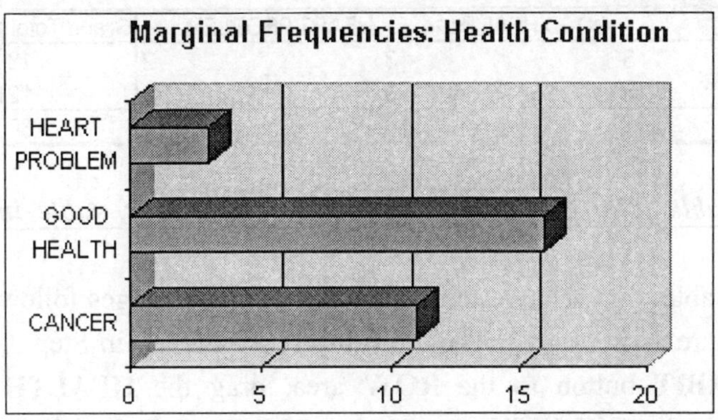

To draw the bar chart for the frequencies in percentage form for HABIT, follow as above and select data in cells **A57:B58**. See Figure 5.14 for the resulting bar chart. Similarly, for HEALTH CONDITION, select data in cells **A67:B69**. Figure 5.15 gives the bar chart for HEALTH CONDITION.

Figure 5.14 Bar chart for relative frequencies - HABIT

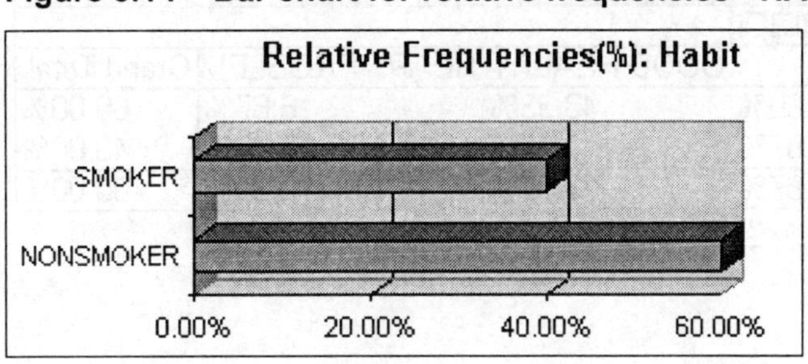

Figure 5.15 Bar chart for relative frequencies - HEALTH CONDITION

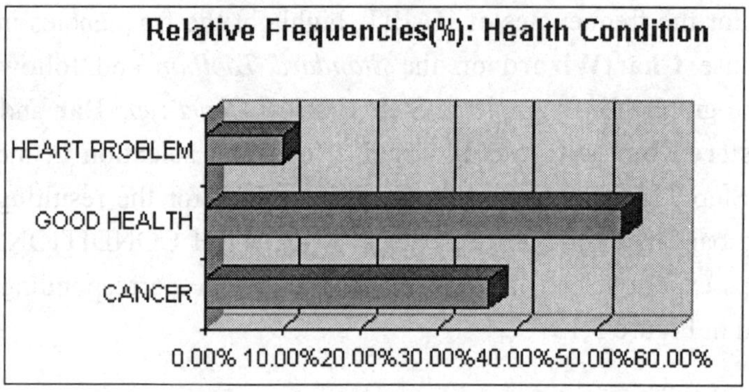

Task 4 ***Constructing two-variable joint frequency table in absolute form***

To construct a two-variable joint frequency table follow the same 4-step procedure as in Task 1. The only differences are that, in Step 3, after dragging the **HABIT** button to the **ROW** area, drag the **HEALTH CONDITION** button to the **COLUMN** area, and in Step 4, select cell **A85** for *PivotTable Starting cell.* See Figure 5.16 for the resulting joint frequency table.

Figure 5.16 Joint frequency table

	A	B	C	D	E
85	Count of ID	HEALTH CONDITION			
86	HABIT	CANCER	GOOD HEALTH	HEART PROBLEM	Grand Total
87	NONSMOKER	3	13	2	18
88	SMOKER	8	3	1	12
89	Grand Total	11	16	3	30

Constructing two-variable joint probability (or relative frequency) table in percentage form

To construct a two-variable joint relative frequency table in percentages follow the same 4-step procedure as in Task 2. The only difference is that, in Step 3, after dragging the **HABIT** button to the **ROW** area, drag the **HEALTH CONDITION** button to the **COLUMN** area, and on the **Pivot Table Field** extended dialogue box instead of *% of column,* now select **% of Total**. In Step 4, select cell **A90** for *PivotTable Starting cell.* See Figure 5.17 for the resulting joint probability table (in percentages).

Figure 5.17 Joint relative frequency table (in percentages)

	A	B	C	D	E
90	Count of ID	HEALTH CONDITION			
91	HABIT	CANCER	GOOD HEALTH	HEART PROBLEM	Grand Total
92	NONSMOKER	10.00%	43.33%	6.67%	60.00%
93	SMOKER	26.67%	10.00%	3.33%	40.00%
94	Grand Total	36.67%	53.33%	10.00%	100.00%

Task 5 ***Drawing a bar chart for the joint relative frequencies***

To draw the bar chart, first we need to edit and copy the joint relative frequency table (Figure 5.17) from Task 4. In cells A100 to A102, type the headings 'HABIT', 'Non-Smoker' and 'Smoker'. Highlight cells **B91** to **D93** and click on **Edit** on the *Menubar* and select **Copy**. Then click the mouse on **B100** and click on **Edit** again on the *Menubar* and select **Paste**. Now highlight cells A100 to D102 and use **ChartWizard** to draw the bar chart (for example, in Step 2, select *Chart type* **Bar;** *Chart sub-type* **Stacked Bar**, the second selection in row 1); in Step 3, give a title for the chart; and in Step 4, click **Finish**). See Figure 5.18 for the bar chart for the joint relative frequencies.

Figure 5.18 Bar chart for joint relative frequencies

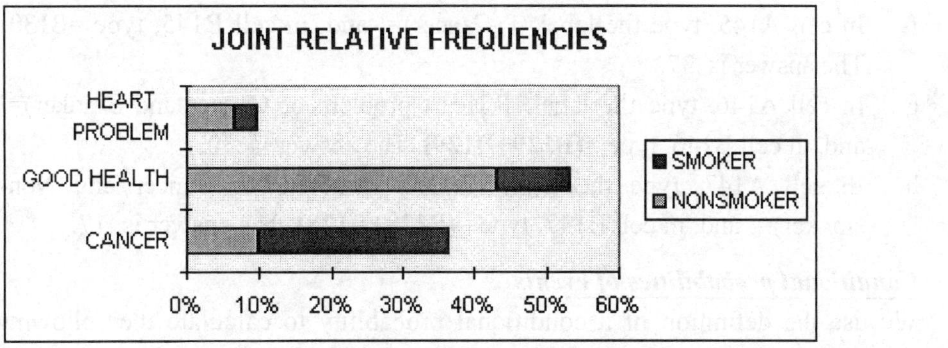

Task 6 ***Converting the joint probability table from percentages to decimals***

As can be seen from Figure 5.17, the probabilities in the joint relative frequency (or probability) table are in percentages. To convert the probabilities from percentages to decimals, we first make a copy of this table into cells A126 to E130. Highlight cells **A90:E94** and click the icon **Copy** on the *Standard Toolbar*. Now click the mouse on cell **A126** and click the icon **Paste** on the *Standard Toolbar*. This will copy cells A90 to E94 onto cells **A126** to **E130**. You can also use the **Copy** and **Paste** options in the **Edit** menu on the *Menubar*. Now highlight cells **B128** to **E130** and click on **Format** on the *Menubar* and select **Cells... Ctrl+1**. A *Format Cells* dialogue box will appear. Select *Category* **Number** and then set decimal places to 2 using the arrows in the dialogue box and click **OK**. Now you will see that all the probabilities are converted to 2 decimals (see Figure 5.19).

Figure 5.19 Joint probability table

	A	B	C	D	E
126	Count of ID	HEALTH CONDITION			
127	HABIT	CANCER	GOOD HEALTH	HEART PROBLEM	Grand Total
128	NONSMOKER	0.10	0.43	0.07	0.60
129	SMOKER	0.27	0.10	0.03	0.40
130	Grand Total	0.37	0.53	0.10	1.00

Joint and marginal probabilities of events

a. In cell A140, type the label 'P(Non-smoker and Heart problem)=' and adjust the column width. In cell B140, type the joint probability **=D128**. The answer is .07.

b. In cell A141, type the label 'P(Smoker and Heart problem)=' and, in cell B141, type the joint probability **=D129**. The answer is .03.

c. In cell A142, type the label 'P(Smoker and Cancer)=' and, in cell B142, type the marginal probability **=B129**. The answer is .27.

d. In cell A143, type the label 'P(Smoker)=' and, in cell B143, type the marginal probability **=E129**. The answer is .40.

e. In cell A144, type the label 'P(Good health)=' and, in cell B144, type =**C130**. The answer is .53.

f. In cell A145, type the label 'P(Cancer)=' and, in cell B145, type **=B130**. The answer is .37.

g. In cell A146, type the label 'P(Heart problem or Cancer and Smoker)=' and, in cell B146, type **=(B129+D129)**. The answer is .30.

h. In cell A147, type the label 'P(Heart problem or Cancer and Non-smoker)=' and, in cell B147, type **=(B128+D128)**. The answer is .17.

Task 7 ***Conditional probabilities of events***

We use the definition of a conditional probability to calculate the following probabilities:

a. $$\text{P(Smoker} \mid \text{Cancer)} = \frac{\text{P(Smoker and Cancer)}}{\text{P(Cancer)}}$$

In cell A150, type 'P(Smoker/Cancer) =' and, in cell B150, type **=B129/B130**. The answer is .73.

b. $$\text{P(Heart problem} \mid \text{Non-smoker)} = \frac{\text{P(Heart problem and Non-smoker)}}{\text{P(Non-smoker)}}$$

In cell A151, type 'P(Heart problem/Non-smoker) =' and, in cell B151, type **=D128/E128**. The answer is .11.

c. $$\text{P(Smoker} \mid \text{Cancer or Heart problem)} = \frac{\text{P(Smoker and Cancer or Heart problem)}}{\text{P(Cancer or Heart problem)}}$$

In cell A152, type 'P(Smoker/Cancer or Heart problem) =' and, in cell B152, type **=(B129+D129)/(B130+D130)**. The answer is .64.

Task 8 Save the workbook with filename **A:EX51.XLS** on your disk and then print the file using the **Print** option on the **File** menu on the *Menubar*.

5.3 Exercises to Chapter 5

Exercise 5.1

The trustee of a mining company's accident compensation plan solicited the employee's feelings toward a proposed revision in the plan. The responses are shown in the following table. There are three employee types (Blue-collar workers, White-collar workers and Managers) and two kinds of decisions (either for or against).

Employee ID	Employee type	Decision
1	Blue collar	For
2	Blue collar	Against
3	Manager	For
4	Blue collar	Against
5	Blue collar	For
6	Blue collar	Against
7	White collar	Against
8	White collar	For
9	Manager	For
10	Blue collar	Against
11	White collar	For
12	Blue collar	For
13	White collar	Against
14	Manager	Against
15	Blue collar	Against
16	White collar	For
17	White collar	For
18	White collar	Against
19	Blue collar	Against
20	Manager	For
21	Blue collar	Against
22	White collar	For
23	Blue collar	Against
24	Manager	Against
25	White collar	For
26	White collar	For
27	White collar	Against
28	White collar	Against
29	Blue collar	For
30	Manager	For

Use EXCEL to perform the following tasks.

Task 1 Construct two separate marginal frequency tables in absolute frequency form for employee types and decision options.

Task 2 Construct two separate marginal relative frequency tables in percentage form for employee types and decision options.

Task 3 Draw bar charts for the above frequencies in absolute and percentage forms.

Task 4 Set up a two-variable joint frequency table in absolute and relative frequency forms.

Task 5 Draw a bar chart for the joint relative frequencies.

Task 6 Save your workbook as 'A:MINING.XLS' on your disk and print the file.

Exercise 5.2

Use the Employee Response survey data given in ***Exercise 5.1*** to answer the following questions using EXCEL.

Task 1 Type the title 'Employees Responses' in cell A1 and in cells A2 to C2 type the column headings 'Employee ID', 'Employee Type' and 'Decision'. Centre the title across columns A1 to C1. Copy the data from Exercise 5.1 file (A:MINING.XLS) using the icon **Copy** and the icon **Paste** available on the *Standard Toolbar* into cells A3 to C32 of this worksheet and save the data as A:EMP.XLS.

Task 2 Set up a joint frequency table for *Employee type* (row) against *Decision* (column).

Task 3 Set up a joint probability table (in percentages) for *Employee type* (row) against *Decision* (column) which would also show the marginal probabilities for *Employee type* and *Decision*.

Task 4 Use the joint probability table from Task 3 to locate/derive the answers to the following questions. If one employee is selected at random,

a. What is the probability that the employee selected is a *blue collar worker* and *for the decision*?

b. What is the probability that the employee selected is a *manager*?

c. What is the probability that the employee selected is *not a manager* and *for the decision*?

d. What is the probability that the employee selected is a *blue collar worker or a manager* and *against the decision*?

e. What is the probability that the employee selected is a *blue collar worker and against the decision* or a *manager and for the decision*?

Task 5 a. Given that the employee selected is *against the decision*, what is the probability that he/she is a *blue collar worker*?

b. If the employee selected is a *manager*, what is the probability that he/she is *for the decision*?

c. If the employee selected is a *blue collar worker* or a *white collar worker*, what is the probability that he/she is *against the decision*?

Task 6 Save the entire EXCEL workbook on your disk as 'A:EMP.XLS' and print the file.

Chapter 6

Discrete Probability Distributions

A random variable is called *discrete* if it can assume a finite number or countably infinite number of numerical values. A *discrete probability distribution* is a table or a graph that assigns each possible value that the discrete random variable can assume together with the associated probability. Using the probability distribution, we can calculate the expected value (mean) and the variance of the random variable. The two important discrete random variables we consider in this chapter are the *binomial* and the *Poisson* random variables.

6.1 Expected value and variance

Let X be a discrete random variable which can assume values x_i, i=1,2,…,*n*, with associated probabilities $p_i = P(X=x_i)$, i=1,…,*n*, respectively. The *expected value* or *mean* of a random variable X, denoted by μ or E[X], is the weighted sum of all possible values of the random variable where the weights are the corresponding probabilities. That is,

$$\mu = E[X] = \sum_{i=1}^{n} x_i p_i$$

The *variance* of X is defined as

$$\sigma^2 = \mathrm{Var}[X] = E\left[(X-\mu)^2\right] = E\left[X^2\right] - \mu^2 = \sum_{i=1}^{n} x_i^2 p_i - \mu^2$$

The *standard deviation* is defined as the square root of the variance.

In EXCEL, the expected value and variance of X can be obtained by using the icon Σ **Autosum** or by writing basic algebraic operations based on the above formulae in a selected cell.

Aim: At the completion of the examples in this chapter, you should be able to use **EXCEL** to calculate the expected value (mean), variance and the standard deviation of a given discrete probability distribution; and calculate various probabilities of a random variable that follows either a Binomial or a Poisson distribution.

6.2 Example 6.1: Calculating the expected value, variance and standard deviation of a random variable

(Ref: Based on Exercise 5.14, p.134, Australian Business Statistics, Selvanathan et al, 1994)

The following table gives the number of times a student visits a nearby pizza shop in a one-month period (denoted by X). The following table is the probability distribution of X (assume that the maximum number of times the student visits the pizza shop does not exceed 3 in one month):

X=x	0	1	2	3
P(X=x)	.1	.3	.4	*p*

Use EXCEL to perform the following tasks.

Task 1 What is the probability that the student visits the pizza shop exactly 3 times a month? What is the probability that the student visits the pizza shop at least once in a month?

Task 2 Draw a histogram of the probability distribution.

Task 3 Calculate the expected value (mean), variance and standard deviation of the random variable X from first principles (i.e., using the definition).

Task 4 If the student buys 2 pizzas per visit at a membership discount rate of $3.50 per pizza (where he pays a monthly membership of $1 to be eligible for the discount price), what is the monthly total cost (Y) in terms of X. Set up the probability distribution table for the random variable Y. Calculate the mean (expected value), variance and standard deviation of Y (the monthly total cost to the student to buy pizza) from first principles. Verify that your results together with the ones in Task 3 satisfy the rules

$$E(aX + b) = aE(X) + b \text{ and } Var(aX + b) = a^2Var(X)$$

for any constants *a* and *b*.

Task 5 Save your EXCEL workbook as 'A:EX61.XLS' on your disk and print the file.

Solution for Example 6.1

Open a new EXCEL workbook. First, type the label 'X=x' in cell A1 and the x values in cells A2 to A5, and the label 'P(X=x)' in cell B1 and the probability values in cells B2 to B4. Leave cell B5 blank at this stage.

Task 1 To calculate the missing probability, P(X=3), assuming that X does not take any value greater than 3, we use the Probability Rule that the sum of all the probabilities for all possible values of X is 1. Therefore,

$$p = P(X=3) = 1 - P(X<3) = 1 - P(X=0) - P(X=1) - P(X=2)$$

In order to use EXCEL to obtain the value of p, in cell B5, type the formula **=1-B2-B3-B4**, and then press **Enter**. You will see the value for p, .2, appearing in cell B5.

We have, $P(X \geq 1) = P(X=1) + P(X=2) + P(X=3)$. Now, to calculate the probability that the student visits the pizza shop at least once a month, in cell **A6,** type the label 'P(X>=1)=' and, in cell **B6,** type the formula **=B3+B4+B5** and press **Enter**.

Task 2 To draw the histogram of the probability distribution, highlight cells **A1** to **B5** and use **ChartWizard** on the *Standard toolbar* and follow Steps 1-4 to draw the graph (see Task 3 of Example 3.2). The corresponding histogram is shown in Figure 6.1.

Figure 6.1 Histogram of the probability distribution

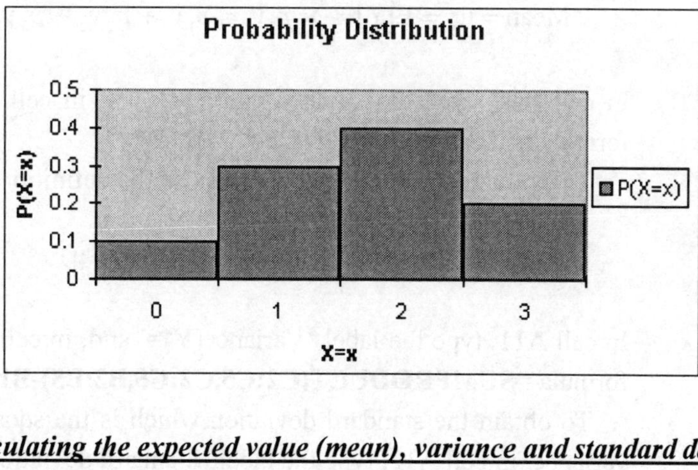

Task 3 ***Calculating the expected value (mean), variance and standard deviation***

To evaluate the expected value (mean) of X, we use

$$\text{Mean} = \mu = E[X] = \sum p_i x_i = p_1 x_1 + p_2 x_2 + p_3 x_3 + p_4 x_4 ,$$

which is the sum of the product of the two columns A2:A5 and B2:B5. In cell A7, type the label 'Mean(X)='. To obtain the sum of the product of the two columns in cell B7, either use the **Paste Function** on the *Standard toolbar,* select option **All,** then select *Function Name* **SUMPRODUCT** and type A2:A5 for Array 1 and type B2:B5 for Array 2, or type the formula **=SUMPRODUCT(A2:A5,B2:B5).**

To evaluate the variance of X, we use

$$\text{Variance} = \sigma^2 = E[X^2] - \mu^2 = \sum x_i^2 p_i - \mu^2 .$$

In cell A8, type the label 'Variance(X)=' and, in cell B8, type the formula **=SUMPRODUCT(A2:A5,A2:A5,B2:B5)-B7*B7**. In cell A9, type the label 'Standard deviation(X)=' and, in cell B9, type the formula **=SQRT(B8).**

Task 4 ***Constructing a probability distribution for Y=aX+b***

The monthly total cost (Y) is $7 times the number of times the student visits the pizza shop per month (X) plus the monthly membership fee of $1. Thus, we can express the monthly total cost Y in terms of X in the following way:

Y=7*X+1

To construct the probability distribution table for Y, in cell C1, type the title 'Y=y' and, in cell C2, type the formula **=7*A2+1**. Then copy this formula to cells C3 to C5.

Method 1 ***Calculating the expected value (mean), variance and standard deviation of Y from first principles***

To evaluate the mean of Y, we use the definition

$$\text{Mean} = \mu_y = E[Y] = \sum p_i y_i = p_1 y_1 + p_2 y_2 + p_3 y_3 + p_4 y_4 .$$

In cell A10, type the label 'Mean(Y)=' and, in cell B10, type the formula **=SUMPRODUCT(C2:C5,B2:B5).**

To evaluate the variance of Y, we use the definition

$$\text{Variance} = \sigma^2 = E[Y^2] - \mu^2 = \sum y_i^2 p_i - \mu^2 .$$

In cell A11, type the label 'Variance(Y)=' and, in cell B11, type the formula **=SUMPRODUCT(C2:C5,C2:C5,B2:B5)-B10*B10.**

To obtain the standard deviation which is the square root of the variance, in cell A12, type the label 'Standard deviation(Y)=' and, in cell B12, type the formula **=SQRT(B11).**

Method 2 ***Calculating the expected value (mean), variance and standard deviation of Y using E(aX+b)=aE(X)+b and Var(aX+b)=a²Var(X)***

In cells A13 and A14 type the labels 'a=' and 'b=' and in cells B13 and B14 type the numbers **7** and **1**. To calculate E(Y)=a*E(X)+b, in cell A15, type the label 'E(Y)='and, in cell B15, type the formula

=B13*B7+B14. To calculate Var(Y)=a^{2}Var(X), in cell A16, type the label 'Var(Y)=' and in cell B16, type the formula **=B13*B13*B8**. To calculate the standard deviation, in cell A17 type the text 'SD(Y)=' and, in cell B17, type the formula **=SQRT(B16)**. Figure 6.2 presents the output.

Figure 6.2 Summary statistics results

	A	B	C
1	X=x	P(X=x)	Y=y
2	0	0.1	1
3	1	0.3	8
4	2	0.4	15
5	3	0.2	22
6	P(X>1)=	0.9	
7	Mean(X)=	1.7	
8	Variance(X)=	0.81	
9	Standard Deviation(X)=	0.9	
10	Mean(Y)=	12.9	
11	Variance(Y)=	39.69	
12	Standard Deviation(Y)=	6.30	
13	a=	7	
14	b=	1	
15	E(Y)=	12.9	
16	Var(Y)=	39.69	
17	SD(Y)=	6.3	

As can be seen from comparing the means in B10 and B15, the variances in B11 and B16, and the standard deviations in B12 and B17, the values are identical. This verifies the rules, E(aX+b)=aE(X)+b and Var(aX+b)=a^2Var(X), for any constants a and b.

Task 5 Save your workbook as **A:EX61.XLS** on your disk and print the file.

6.3 Binomial distribution

If a random experiment has only two possible outcomes, then it is called a *Bernoulli* trial. The two possible outcomes of a Bernoulli trial are usually called 'success' and 'failure'. A *binomial experiment* consists of n independent Bernoulli trials, where the probability of obtaining a success in any single trial is a constant and is equal to p. The random variable that counts the total number of successes in a binomial experiment with n trials is called a *binomial random variable*. The binomial random variable is a discrete random variable which can take any one of the values, 0, 1, 2, …, n, and its probability distribution is called a *binomial distribution.* A simple example of a binomial experiment is flipping a coin 100 times and counting the number of heads (successes). The random variable X that records the number of heads (successes) in the n trials is a binomial random variable and the probability distribution of X is a binomial distribution.

A binomial distribution is characterised by two parameters, namely, the total number of trials (n) in the experiment and the probability of a success (p). The expected value and the variance of a binomial random variable is given by

$$E[X] = np; \qquad Var[X] = npq, \qquad \text{where } q=1-p.$$

In EXCEL, to calculate the binomial probabilities, we first click on the icon **Paste Function** on the *Standard toolbar* (see Figure 6.3) and then select the *Function Category* either **All** or **Statistical** and *Function Name* **BINOMDIST**. *Alternatively*, we can use the following formulae depending on the type of probability is required.

$$
\begin{aligned}
P[X=x] &= \text{BINOMDIST}(x,n,p,\text{FALSE}) \\
P[X \leq x] &= \text{BINOMDIST}(x,n,p,\text{TRUE}) \\
P[x \leq X \leq y] &= P(X \leq y)-P(X \leq x-1) \\
&= \text{BINOMDIST}(y,n,p,\text{TRUE}) - \text{BINOMDIST}(x-1,n,p,\text{TRUE})
\end{aligned}
$$

Figure 6.3 Statistical Function categories

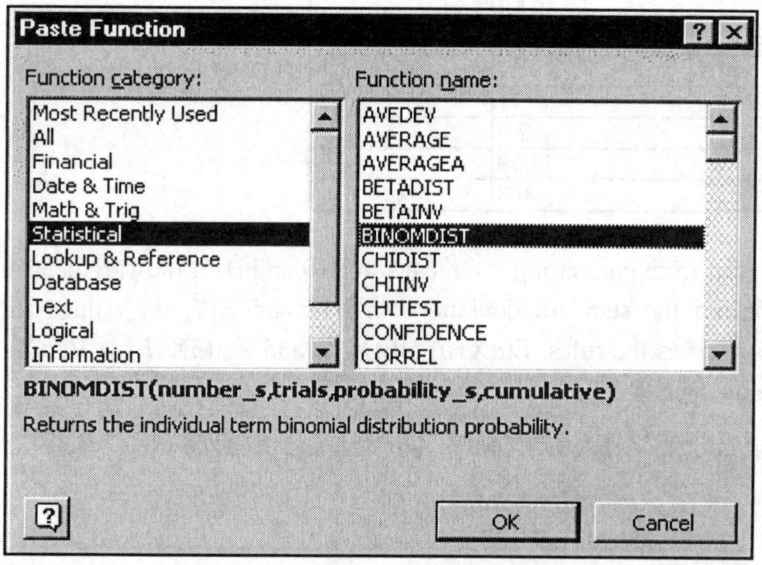

For example, if X~ Binominal (n=10, p=.2), to obtain $P[X \leq 3]$, type the formula **=BINOMDIST(3,10,.2,TRUE)**.

6.4 Example 6.2: Calculating the probabilities, mean, variance and standard deviation of a binomial random variable

A car yard sales record shows that 40 percent of the customers use cheques as a medium for their payment. It is expected that following one of their clearance sales advertisement on the TV, 10 customers will purchase cars on the following day. Let X be the number of these customers who pay by cheque. Use EXCEL to find the probabilities for the events in Tasks 1-8, and perform Tasks 9-12.

Task 1 P(At most 3 customers make their payment by cheque)
Task 2 P(Exactly 3 customers make their payment by cheque)
Task 3 P(At least 6 customers make their payment by cheque)
Task 4 P(More than 8 customers make their payment by cheque)
Task 5 P(Less than 7 customers make their payment by cheque)
Task 6 P(More than 4 and less than or equal to 8 customers make their payment by cheque)
Task 7 P(Between 7 and 9 customers, both inclusive, make their payment by cheque)
Task 8 P(At least 6 or more and less than 9 customers make their payment by cheque)
Task 9 Construct the complete probability distribution of X (= the number of customers who make their payment by cheque). Plot these probabilities in a histogram.
Task 10 From the probability distribution, calculate the mean, variance and standard deviation of X (= the number customers making their payment by cheque).
Task 11 Verify that your answers from Task 10 are the same as those obtained from formulae Mean=*np*, Variance=*npq* and Standard Deviation=SQRT(*npq*).
Task 12 Save your EXCEL workbook as 'A:EX62.XLS' on your disk and print the file.

Solution to Example 6.2

It can be easily seen that X (the number of customers who make their payment by cheque) can be considered as a ***binomial random variable*** with parameters n = Total number of customers who purchase the car=10; and p=P[a customer making the payment by cheque]=.40.

To calculate the binomial probabilities using EXCEL, we use the icon **Paste Function** on the *Standard toolbar* or type the formula in the cell. To use the formula, in general, if **X ~ Binomial(*n,p*)**, then for any value x=0,1,2,....,n, to obtain the cumulative probability **P(X≤x)**, type the formula **=BINOMDIST(x,*n,p*,TRUE)** and to obtain the individual probability **P(X=x)**, type the formula **=BINOMDIST(x,*n,p*,FALSE)**.

Task 1 To calculate the probability that at most 3 customers make their payment by cheque, i.e., P(X≤3), in cell A1, type the label 'P(X<=3)='. Now click the mouse on cell B1.

Step 1 Click on the icon **Paste Function** on the *Standard toolbar* and then from the dialogue box, select *Function Category* either **All** or **Statistical.** Then select *Function Name* **BINOMDIST** and click **OK**.

Step 2 Now the BINOMDIST dialogue box will appear and you will be required to input four values (see Figure 6.4).

Figure 6.4 BINOMDIST Function dialogue box

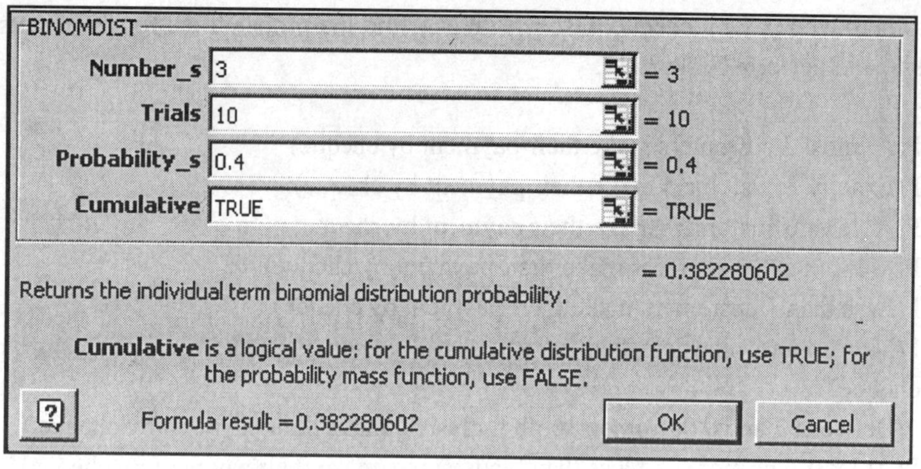

1. *Number s* - the value of X (for this task enter 3)
2. *Trials* - the number of trials *n* (for this task enter 10)
3. *Probability s* - the probability of a success *p* (for this task enter .4)
4. *Cumulative* - Enter TRUE if you require a cumulative probability P(X≤3) or FALSE if you require the exact probability P(X=3). For this task, as we require P(X≤3), enter **TRUE**. Then click on **OK**.

Now you will see the value of P(X≤3) in cell B1.

Alternatively, to obtain P(X≤3) in cell B1, simply type the formula **=BINOMDIST(3,10,.4,TRUE)**. [On the other hand, if you require P(X=3), type the formula **=BINOMDIST (3,10,.4,FALSE)**].

Task 2 To calculate the probability that exactly 3 customers pay by cheque, i.e. P(X=3), first in cell A2, type label 'P(X=3)=' and then follow Steps 1 and 2 as for Task 1 except in Step 2 where you enter **FALSE** for the fourth entry.

Alternatively, in cell B2, type the formula **=BINOMDIST(3,10,.4,FALSE)**, to obtain P(X=3).

Task 3 To calculate the probability that at least 6 customers pay by cheque, i.e. P(X≥6), in cell A3, type the label 'P(X>=6)='. Note that P(X≥6)=1-P(X≤5). Therefore, to obtain P(X≥6), in cell B3, type the formula **=1-BINOMDIST(5,10,.4,TRUE)**.

Task 4 To calculate the probability that more than 8 customers pay by cheque, i.e. P(X>8), in cell A4, type the label 'P(X>8)='. Note that P(X>8)=1-P(X≤8). Therefore, to obtain P(X>8), in cell B4, type the formula **=1-BINOMDIST(8,10,.4,TRUE)**.

Task 5 To calculate the probability that less than 7 customers pay by cheque, i.e. P(X<7), in cell A5, type the label 'P(X<7)='. Note that P(X<7)=P(X≤6). Therefore, to obtain P(X<7), in cell B5, type the formula **=BINOMDIST(6,10,.4,TRUE)**.

Task 6 To calculate the probability that more than 4 and less than or equal to 8 customers pay by cheque, i.e. $P(4<X\leq 8)$, in cell A6, type the label 'P(4<X<=8)='. Note that $P(4<X\leq 8)=P(5\leq X\leq 8)=P(X\leq 8)-P(X\leq 4)$. Therefore, to obtain $P(4<X\leq 8)$, in cell B6, type the formula **=BINOMDIST(8,10,.4,TRUE)-BINOMDIST(4,10,.4,TRUE)**.

Task 7 To calculate the probability that between 7 and 9 customers (both inclusive) pay by cheque, i.e. $P(7\leq X\leq 9)$, in cell A7, type the label 'P(7<=X<=9)='. Note that $P(7\leq X\leq 9)=P(X\leq 9)-P(X\leq 6)$. Therefore, to obtain $P(7\leq X\leq 9)$, in cell B7, type the formula **=BINOMDIST(9,10,.4,TRUE)-BINOMDIST(6,10,.4,TRUE)**.

Task 8 To calculate the probability that at least 6 or more and less than 9 customers pay by cheque, i.e. $(P(6\leq X<9)$, in cell A8, type the label 'P(6<=X<9)='. Note that $P(6\leq X<9)=P(6\leq X\leq 8)=P(X\leq 8)-P(X\leq 5)$. Therefore, to obtain $P(6\leq X<9)$, in cell B8, type the formula **=BINOMDIST(8,10,.4,TRUE)-BINOMDIST(5,10,.4,TRUE)**.

The output from Tasks 1-8 can be seen in Figure 6.5.

Figure 6.5 Binomial probabilities and distribution

	A	B	C	D	E	F
1	P(X<=3)=	0.3823				
2	P(X=3)=	0.2150				
3	P(X>=6)=	0.1662				
4	P(X>8)=	0.0017				
5	P(X<7)=	0.9452				
6	P(4<X<=8)=	0.3652				
7	P(7<=X<=9)=	0.0547				
8	P(6<=X<9)=	0.1646				
9	X=x	P(X=x)				
10	0	0.0060				
11	1	0.0403				
12	2	0.1209				
13	3	0.2150				
14	4	0.2508				
15	5	0.2007				
16	6	0.1115				
17	7	0.0425				
18	8	0.0106				
19	9	0.0016				
20	10	0.0001				
21		1				
22						
23						
24		From First principle:	Using Formula			
25	Mean =	4	4			
26	Variance=	2.4	2.4			
27	Standard deviation=	1.549193338	1.549193338			

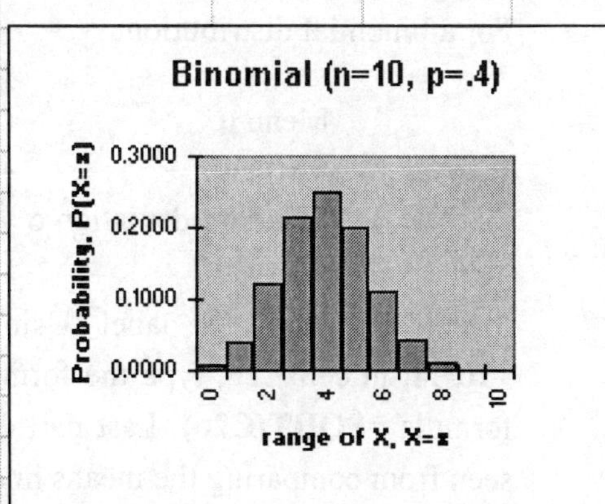

Task 9 ***Constructing a complete binomial probability distribution***

To construct the complete Binomial(n=10,p=.4) distribution, we need to obtain the probabilities for X = 0,1,...,10; that is, P(X=0), P(X=1), ..., P(X=10).

In cells A9 and B9, type the headings 'X=x' and 'P(X=x)'. In cell A10, type **0** and in cell B10, type the formula **=BINOMDIST(A10,10,.4,FALSE)**. In cell A11, type the formula **=A10+1**. Copy this formula into cells A12 to A20 using the pointer +. [Otherwise, type **1** in cell A11**,** highlight cells A10 and A11, and using the pointer + drag down up to cell A20.] Now copy the formula in cell B10 into cells B11 to B20 using the pointer +. This will complete the values for the X=x and P(X=x) columns. As a check highlight all values in cells B10 to B20 and use the icon Σ **Autosum** on the *Standard toolbar* to see whether the probabilities add up to 1.

Now use **ChartWizard** to draw the probability distribution in the form of a histogram (see Chapter 3). The final output can be seen in Figure 6.5.

Task 10 ***Calculating the expected value (mean), variance and standard deviation of X from first principles using the individual probability values***

In cell B24, type the heading 'From First Principles'. To calculate the mean $\mu=E[X]=\sum x_i p_i$, in cell A25, type the label 'Mean=' and, in cell B25, type the formula **=SUMPRODUCT(A10:A20, B10:B20)**.

To calculate the variance of $X, \sigma^2 = E[X^2] - \mu^2 = \sum x_i^2 p_i - \mu^2$, in cell A26, type the label 'Variance=' and, in cell B26, type the formula **=SUMPRODUCT(A10:A20,A10:A20,B10:B20)-B25*B25**. To obtain the standard deviation of X, σ, in cell A27, type the label 'Standard deviation=' and, in cell B27, type the formula **=SQRT(B26)**.

Task 11 ***Calculating the expected value (mean), variance and standard deviation of the binomial random variable X using the formulae***

For a binomial distribution,

$$\begin{aligned} \text{Mean } \mu &= np \\ \text{Variance } \sigma^2 &= npq \\ \text{Standard deviation } \sigma &= \sqrt{npq}\,. \end{aligned}$$

In cell C24, type the label 'Using Formula'. In cell C25, type the formula **=10*.4**; in cell C26, type the formula **=10*.4*(1-.4)**; and in cell C27, type the formula **=SQRT(C26)**. Last part of Figure 6.5 presents the output. As can be seen from comparing the means in cells B25 and C25, the variances in cells B26 and C26 and the standard deviations in cells B27 and C27, we see that the values are identical.

Task 12 Save your EXCEL workbook as **A:EX62.XLS** on your disk and print the file.

6.5 Poisson distribution

While the binomial random variable counts the number of successes that occur in a fixed number of trials, a *Poisson random variable X* counts the number of rare events that occur (successes) in a specified time interval or a specified region. For example, the number of typing errors on a page of a book is a Poisson random variable. Therefore the Poisson random variable *X* is a discrete random variable that can take values 0, 1, 2, … The *Poisson distribution* is characterised by a single parameter, the average number of successes occurring in the given interval (μ).

Using EXCEL, to calculate the Poisson probabilities, we use the same procedure as for the binomial probabilities. We first click on the **Paste Function** on the *Standard toolbar* (see Figure 6.1) and then select the *Function Category* either **All** or **Statistical** and *Function Name* **POISSON**. *Alternatively*, we can use the following formulae.

$P[X=x]$ = POISSON(x,μ,FALSE)
$P[X \leq x]$ = POISSON(x,μ,TRUE)
$P[x \leq X \leq y]$ = $P[X \leq y]-P[X \leq x-1]$
= POISSON(y,μ,TRUE)-POISSON($x-1$,μ,TRUE)

For example, if X has a Poisson distribution with average μ=2. to obtain $P[X \leq 3]$, type the formula =**POISSON (3,2,TRUE)**.

6.6 Example 6.3: Calculating the Poisson probabilities

The marketing manager of a mail order company has noted that, on average, she receives 10 complaint calls from customers during a week (consisting of 5 working days), and that the calls occur at random. Using EXCEL, obtain the probabilities for the following events in Tasks 1-8, and perform Tasks 9-12.

Task 1 P(At most 12 complaint calls *in a week*)
Task 2 P(Exactly 9 complaint calls *in a week*)
Task 3 P(At least 8 complaint calls *in a week*)
Task 4 P(More than 2 complaint calls *in a single day*)
Task 5 P(Less than 3 complaint calls *in a single day*)
Task 6 P(More than 1 and less than or equal to 3 complaint calls *in a single day*)
Task 7 P(Between 1 and 4 complaint calls, both inclusive, *in a single day*)
Task 8 P(Exactly 5 complaint calls *in a single day*)
Task 9 Construct the complete probability distribution of Y (= the number of complaint calls *in a single day*) for Y=0,1,2,….,15. Plot these probabilities in a histogram.
Task 10 From the probability distribution, calculate the mean, variance and standard deviation of Y (= the number of complaint calls *in a single day*).

Task 11 Verify that your answers from Task 10 are approximately the same as those obtained from the formulae Mean=μ_l, Variance=μ_l and Standard deviation=SQRT(μ_l), where μ_l is the average number of complaint calls received *in a single day*.

Task 12 Save your EXCEL workbook as 'A:EX63.XLS' on your disk and print the file.

Solution to Example 6.3

Let X be the number of complaint calls **received during a week**. Then X can be considered as a ***Poisson random variable*** with parameter μ=average number of complaint calls received during a week=10.

To calculate the Poisson probabilities using EXCEL, we use the icon **Paste Function** on the *Standard toolbar*, or type the formula in the cell. If using the formula, in general, if **X ~ Poisson(μ)**, then for any value **x=0,1,2,....** to obtain the cumulative probability **P(X≤x)**, type the formula **=POISSON(x,μ,TRUE)** and to obtain the individual probability **P(X=x)**, type the formula **=POISSON(x,μ,FALSE)**.

Task 1 Open a new EXCEL workbook. To obtain the probability of receiving at most 12 complaint calls in a week, i.e. (P(X≤12), in cell A1, type the label 'P(X<=12)='. Now click the mouse on cell B1.

<u>Step 1</u> Click on the icon **Paste Function** on the *Standard toolbar* and then from the dialogue box, select *Function Category* either **All** or **Statistical.** Then select *Function Name* **POISSON** and click **OK**.

<u>Step 2</u> Now the POISSON dialogue box will appear and you will be required to input three values (see Figure 6.6).

Figure 6.6 POISSON Function dialogue box

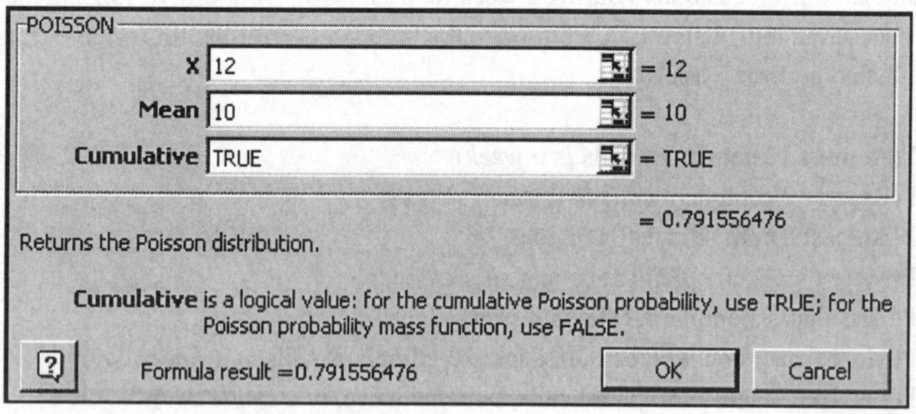

1. <u>*x*</u> - the value of X. For example, if we need P(X=12) or P(X≤12), enter 12.
2. <u>*Mean*</u> - the mean of the Poisson distribution (μ). For this example, enter 10.
3. <u>*Cumulative*</u> - Enter TRUE if you require a cumulative probability P(X≤12) or enter FALSE if you require a probability such as P(X=12). As we require P(X≤12), enter **TRUE**. Then click **Finish**.

Now you will see the value of P(X ≤ 12) in cell B1.

Alternatively, to obtain P(X≤12) in cell B1, simply type the formula **=POISSON(12,10,TRUE)**. [On the other hand, if you require P(X=12), type the formula **=POISSON(12,10,FALSE)**.]

Task 2 To obtain P(X=9), first in cell A2, type label 'P(X=9)=' and then follow Steps 1 and 2 as for Task 1 except in Step 2 enter '**FALSE**' for the third entry.

Alternatively, to obtain P(X=9) in cell B2, type the formula **=POISSON(9,10,FALSE)**.

Task 3 To obtain (P(X≥ 8), in cell A3, type the label 'P(X>=8)='. Note that P(X≥ 8)=1-P(X≤7). Therefore, in cell B3, type the formula **=1-POISSON(7,10,TRUE)**.

Task 4 **Let Y be the number of complaint calls received during a single day.** As one week consists of 5 working days, μ_l=average number of complaint calls received during a day=10/5=2. Therefore, the random variable Y has a ***Poisson distribution*** with mean μ_l=2.

To obtain P(Y>2), in cell A4, type the label 'P(Y>2)='. Note that P(Y>2)=1-P(Y≤2). Therefore, in cell B4, type the formula **=1-POISSON(2,2,TRUE)**.

Task 5 To obtain P(X<3), in cell A5, type the text, 'P(X<3)='. Note that P(X<3)=P(X≤2). Therefore, in cell B5, type the formula = **POISSON (2,2,TRUE)**.

Task 6 To obtain P(1<Y≤3) in cell A6, type the label 'P(1<Y<=3)='. Note that P(1<Y≤3)=P(2≤Y≤3)=P(Y≤3)-P(Y≤1). Therefore, in cell B6, type the formula **=POISSON(3,2,TRUE)-POISSON(1,2,TRUE)**.

Task 7 To obtain P(1≤Y≤4), in cell A7, type the label 'P(1<=Y<=4)='. Note that P(1≤Y≤4)=P(Y≤4)-P(Y≤0). Therefore, in cell B7, type the formula **=POISSON(4,2,TRUE)-POISSON(0,2,TRUE)**.

Task 8 To obtain P(Y=5), in cell A8, type the label 'P(Y=5)='. In cell B8, type the formula **=POISSON(5,2,FALSE)**.

Task 9 ***Constructing a Poisson probability distribution***

To construct the complete Poisson (μ=2) distribution for Y=0,1,...15, we need to obtain the probabilities P(Y=0), P(Y=1), …, P(Y=15).

In cells A9 and B9, type the headings 'Y=y' and 'P(Y=y)'. In cell A10, type **0**. In cell A11, type the formula **=A10+1**. Copy this formula into cells A12 to A25 using the pointer +. In cell B10, type the formula **=POISSON(A10,2,FALSE)** and copy the formula in cell B10 into cells B11 to B25 using the pointer +. This will complete the values for the Y=y and P(Y=y) columns.

Now use **ChartWizard** to draw the probability distribution (see Chapter 3). The final output for Tasks 1-9 are given in Figure 6.7.

Task 10 ***Calculating the expected value (mean), variance and standard deviation of Y using the individual probability values***

In cell B27, type the title 'From first principles'. To calculate the mean $\mu_1 = E[Y] = \sum y_i p_i$, in cell A28, type the label 'Mean=' and, in cell B28, type the formula **=SUMPRODUCT(A10:A25, B10:B25)**.

To calculate the variance of Y, $\sigma_1^2 = E[Y^2] - \mu_1^2 = \sum y_i^2 p_i - \mu_1^2$, in cell A29, type the label 'Variance=' and in cell B29 type the formula **=SUMPRODUCT(A10:A25,A10:A25,B10:B25)-B28*B28**. To obtain the standard deviation of Y, σ_1, in cell A30, type the label 'Standard deviation=' and, in cell B30, type the formula **=SQRT(B29)**. See Figure 6.7 for the output.

Figure 6.7 Poisson probabilities and distribution

	A	B	C	D	E	F
1	P(X<=12)=	0.7916				
2	P(X=9)=	0.1251				
3	P(X>=8)=	0.7798				
4	P(Y>2)=	0.3233				
5	P(X<3)=	0.6767				
6	P(1<Y<=3)=	0.4511				
7	P(1<=Y<=4)=	0.8120				
8	P(Y=5)=	0.0361				
9	Y=y	P(Y=y)				
10	0	0.1353				
11	1	0.2707				
12	2	0.2707				
13	3	0.1804				
14	4	0.0902				
15	5	0.0361				
16	6	0.0120				
17	7	0.0034				
18	8	0.0009				
19	9	0.0002				
20	10	0.0000				
21	11	0.0000				
22	12	0.0000				
23	13	0.0000				
24	14	0.0000				
25	15	0.0000				
26		1.0000				
27		From first Priciples	Using formula			
28	Mean=	2.0000	2			
29	variance=	2.0000	2			
30	Standard deviation=	1.4142	1.4142			

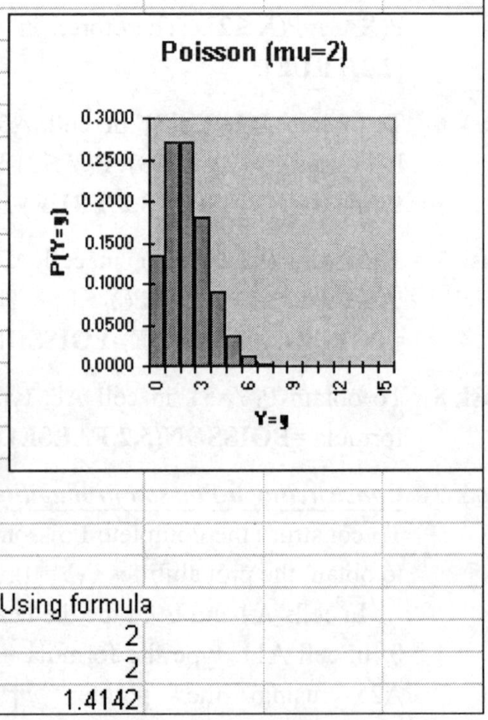

Task 11 ***Calculating the expected value (mean), variance and standard deviation of the Poisson random variable Y using the formulae***

For a Poisson distribution,

$$\text{Mean} = \mu_1$$
$$\text{Variance} = \mu_1$$
$$\text{Standard Deviation} = \sqrt{\mu_1}$$

In cell C27, type text 'Using Formula'. In cell C28, type **2**; in cell C29, type **2**; and in cell C30, type the formula **=SQRT(C29)**. See Figure 6.7 for the output. As can be seen, comparing the means in cells B28 and C28, the variances in cells B29 and C29 and the standard deviations in cells B30 and C30, the values are identical.

Task 12 Save your EXCEL workbook as **A:EX63.XLS** on your disk and print the file.

6.7 Exercises to Chapter 6

Exercise 6.1

Suppose you make a \$1000 investment in a risky venture to make a payoff of X dollars. There are *three* possible outcomes attached to this investment return. *One* is that there is a 40% chance that the payoff from the investment will be \$2000; the *second* is that, a 50% chance that you will just get your money back, and the *third* is that you will receive nothing at all from your investment. Use EXCEL to answer the following tasks.

Task 1 What is the probability that you will receive nothing at all from your investment? What is the probability that you will be able to make at least \$1000 payoff from the investment?

Task 2 Draw a histogram of the probability distribution.

Task 3 Calculate the expected value (mean), variance and standard deviation of the payoff from your investment of \$1000.

Task 4 The net profit (Y) you will make from your investment in terms of payoff X is given by Y=X-1000. Set up the probability distribution table for the random variable Y. Calculate the expected value (mean), variance and standard deviation of your net profit from first principles. Verify that your results together with the ones in Task 3 satisfy the rules $E(aX+b)=a\ E(X)+b$ and $Var(aX+b)=a^2Var(X)$, for any constants a and b.

Task 5 If you invest \$3000 in the risky venture instead of \$1000 and the possible payoffs triple accordingly, what is the expected value (mean), variance and standard deviation of your net profit, from the \$3000 investment?

Task 6 Save the workbook as 'A:RISK.XLS' on your disk and print the file.

Exercise 6.2

The *Financial Magazine* has reported that 80% of firms that speculate on corporate takeovers before they become public also help to underwrite such deals. A random sample of 10 firms that speculate on corporate takeovers is selected. Find the probabilities for the following events in Tasks 1-8 using EXCEL.

Task 1 P(At most half of the firms selected underwrite takeovers)

Task 2 P(At least half of the firms selected underwrite takeovers)

Task 3 P(Exactly half of the firms selected underwrite takeovers)

Task 4 P(At most 3 of the firms selected underwrite takeovers)

Task 5 P(Less than 7 of the firms selected underwrite takeovers)

Task 6 P(More than 3 and less than or equal to 7 of the firms selected underwrite takeovers)

Task 7 P(Between 6 and 8, both inclusive, of the firms selected underwrite takeovers)

Task 8 P(5 or More and less than 8 of the firms selected underwrite takeovers)

Task 9 Construct the complete probability distribution of X (= the number of the firms selected underwrite takeovers). Plot these probabilities in a histogram.

Task 10 From the probability distribution, calculate the mean, variance and standard deviation of X (= the number of the firms selected underwrite takeovers).

Task 11 Verify that your answers from Task 10 are the same as those obtained from the formulae Mean=*np*, Variance=*npq* and Standard Deviation=SQRT(*npq*).

Task 12 Save the EXCEL workbook as 'A:TAKEOV.XLS' on your disk and print the file.

Exercise 6.3

Records show that there is an average of 4 accidents each day within Queensland main city boundaries between 2 and 3pm. Using EXCEL find the probabilities for the events in Tasks 1-13.

Task 1 P(There will be at most two accidents between 2 and 3pm on a particular day)

Task 2 P(There will be exactly two accidents between 2 and 3pm on a particular day)

Task 3 P(There will be at least two accidents between 2 and 3pm on a particular day)

Task 4 P(There will be more than two accidents between 2 and 3pm on a particular day)

Task 5 P(There will be less than two accidents between 2 and 3pm on a particular day)

Task 6 P(There will be at most one accident between 2 and 2.30pm on a particular day)

Task 7 P(There will be exactly one accident between 2 and 2.30pm on a particular day)

Task 8 P(There will be at least one accident between 2 and 2.30pm on a particular day)

Task 9 P(There will be more than two accidents between 2 and 2.30pm on a particular day)

Task 10 P(Between 1 and 4 accidents (both inclusive) between 2 and 2.30pm on a particular day)

Task 11 P(Exactly 5 accidents between 2 and 2.30pm on a particular day)

Task 12 Construct the complete probability distribution of Y (=the number of accidents between 2 and 2.30pm on a particular day) for Y=1,2,….,10. Plot these probabilities in a histogram.

Task 13 From the probability distribution, calculate the mean, variance and standard distribution of Y (=the number of accidents between 2 and 2.30pm on a particular day).

Task 14 Verify that your answers from Task 13 are approximately the same as those obtained from formulae Mean=μ_l, Variance=μ_l and Standard Deviation=SQRT(μ_l), where μ_l is the average number of complaint calls received in a single day.

Task 15 Construct the complete probability distribution of X (=the number of accidents between 2 and 3pm on a particular day) for X=1,2,….,10. Plot these probabilities in a histogram.

Task 16 From the probability distribution, calculate the mean, variance and standard distribution of X (=the number of accidents between 2 and 3pm on a particular day).

Task 17 Verify that your answers from Task 16 are approximately the same as those obtained from formulae Mean=μ, Variance=μ and Standard Deviation=SQRT(μ), where μ is the average number of accidents between 2 and 3pm on a particular day.

Task 18 Save the workbook as ‘A:ACCID.XLS’ on your disk and print the file.

Chapter 7

Continuous Probability Distributions

A random variable is continuous if it can take any value in an interval. (Note that the number of points in an interval is infinite.) For a continuous random variable X, the probability of X taking a specific value, P(X = x), is zero. Therefore, usually, only cumulative probabilities such as $P(X \leq x)$ are required. Also note that, for a continuous probability distribution, $P(X<x)=P(X \leq x)$. Furthermore, $P(X \geq x)=1-P(X \leq x)$.

For a continuous distribution, instead of the probability P(X = x) we consider a function *f(x)* called the *probability density function*. The graph of the probability density function approximates the relative frequency polygon for the population. Some of the well-known continuous probability distributions include the uniform distribution, the normal distribution and the exponential distribution. The most commonly used continuous distribution is the normal distribution.

7.1 Normal distribution

The *normal distribution* is characterised by two parameters, namely the mean (μ) and variance (σ^2). The probability density function of a normal random variable X is

$$f(x)=\frac{1}{\sqrt{2\pi}\sigma}e^{-\frac{1}{2}\left(\frac{x-\mu}{\sigma}\right)^2}, \quad -\infty \leq x < \infty$$

where π = 3.14159, e = 2.71828. The distribution of a normal distribution is bell-shaped and symmetric about the mean μ.

Using EXCEL, to calculate the normal probabilities, we use the same procedure as for the binomial and Poisson probabilities. We first click on the **Paste Function** on the *Standard toolbar* (see Figure 7.1) and then select the *Function Category* either **All** or

Statistical and *Function Name* **NORMDIST**. Alternatively, if X ~ Normal(μ, σ^2) we can use the following formulas.

f(x) = NORMDIST(x,μ,σ,FALSE)
P[X $\leq x$] = NORMDIST(x,μ,σ,TRUE)
P[X $\geq x$] = 1-NORMDIST(x,μ,σ,TRUE)
P[$x \leq$ X $\leq y$] = NORMDIST(y,μ,σ,TRUE)-NORMDIST(x,μ,σ,TRUE)

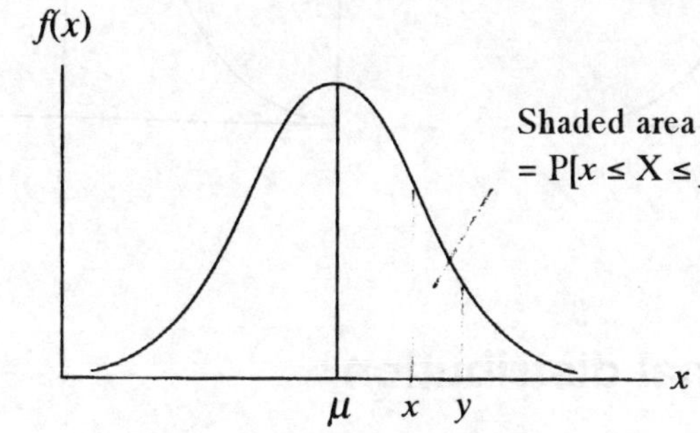

Figure 7.1 Statistical Function categories

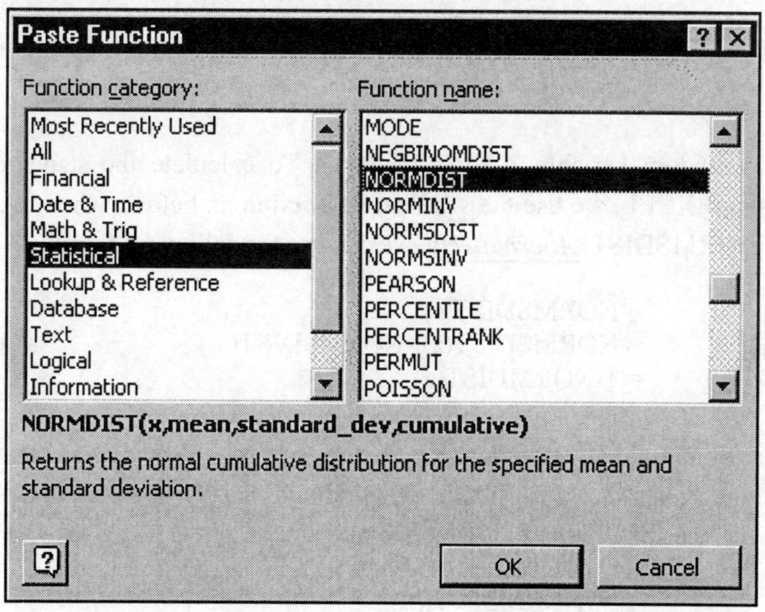

We can also use EXCEL to calculate the x-values corresponding to the cumulative normal probabilities. That is, for a given value of p_o, if P[X $\leq x_o$] = p_o, to obtain x_o we use the **Paste Function** as before except for the *Function Name,* instead of NORMDIST we use **NORMINV**. Alternatively, we can use the following formula.

$$x_o = \text{NORMINV}(p_o,\mu,\sigma)$$

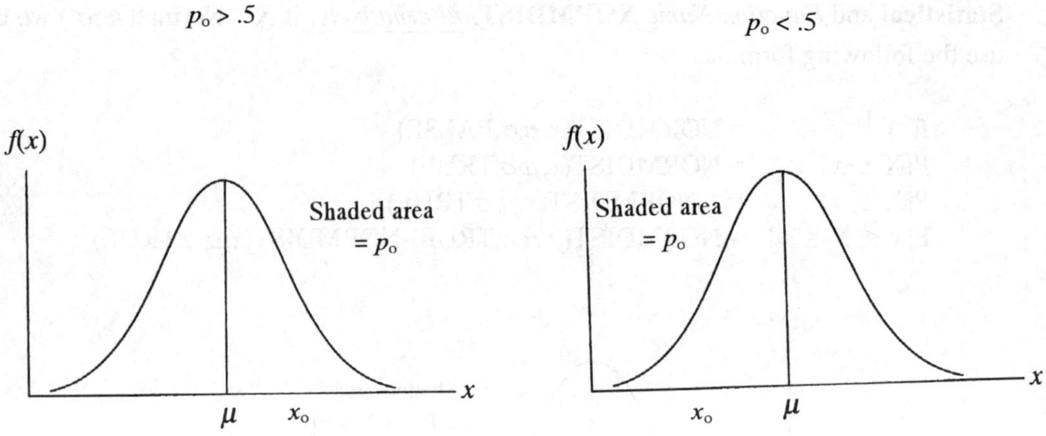

7.2 Standard normal distribution

The normal distribution with zero mean and unit standard deviation is called a **standard normal distribution**. If X is a normal random variable with mean μ and variance σ^2, then the standard normal variable Z is the standardised value of X defined as

$$Z = \frac{X - \mu}{\sigma}$$

The standard normal random variable Z ~ Normal(0,1). To calculate the standard normal probabilities using EXCEL, we use the fx **Paste Function** as before, where we use the *Function Name* NORMSDIST. <u>*Alternatively*</u>, we can use the following formula.

$P[Z \leq z]$	= NORMSDIST(z)
$P[z_1 \leq Z \leq z_2]$	= NORMSDIST(z_2)-NORMSDIST(z_1)
$P[Z \geq z]$	= 1-NORMDIST(z)

Similarly, to calculate the z-value corresponding to a given probability p_o such that $P[Z \leq z_o] = p_o$, we use the *Function Name* **NORMSINV.** [Note that p_o, is a cumulative probability.] <u>*Alternatively*</u>, we can use the following formula.

$$z_o = \text{NORMSINV}(p_o)$$

Aim: At the completion of the examples in this chapter you should be able to use EXCEL to calculate various normal probabilities and the x-values corresponding to given normal probabilities, and graph a normal distribution when its mean and standard deviation are given.

7.3 Example 7.1: Calculating the normal probabilities

(Ref: Based on Exercise 5.54, p. 170, Selvanathan et al, Australian Business Statistics 1994).

A marketing manager of a leading firm believes that total sales for the firm next year can be modelled by using a normal distribution, with a mean of \$2.5m and a standard deviation of \$300,000. Use EXCEL to perform the following tasks.

Task 1 What is the probability that the firm's sales will exceed \$3m?

Task 2 What is the probability that the firm's sales will fall within \$150,000 of the expected level of sales?

Task 3 In order to cover fixed costs, the firm's sales must exceed the break-even level of \$1.8m. What is the probability that sales will exceed the break-even level?

Task 4 Determine the sales level that has only a 9 percent chance of being exceeded next year.

Task 5 Save the EXCEL workbook on your disk as 'A:EX71.XLS' and print the file.

Solution to Example 7.1

Let X be the total sales (in millions of dollars). Then X is normally distributed with mean $\mu = 2.5$ and variance $\sigma^2 = .3^2$. That is, X ~ Normal(2.5,.3^2). [Note that the standard deviation=\$300,000=\$.3 million. Hence the variance $\sigma^2 = .3^2$.] To calculate the normal probabilities using EXCEL, we use the icon **Paste Function** on the *Standard toolbar* or type the formula in the cell as discussed in Section 7.1.

Task 1 To calculate the probability that sales will exceed \$3m, we need P(X>3). As P(X>3) = 1-P(X$\leq$3), we first calculate P(X$\leq$3). In cell A1, type the label 'P(X<=3)='. Now click on cell B1.

Step 1 Click on the icon **Paste Function** on the *Standard toolbar* and then from the dialogue box, select *Function Category* either **All** or **Statistical.** Then select *Function Name* **NORMDIST** and click **OK** on the dialogue box.

Step 2 Now the NORMDIST dialogue box will appear and you will be required to input four values (see Figure 7.2).

1. *x* - the value of X (for this task enter **3**)
2. *Mean* - the mean of the distribution of X (for this task enter **2.5**)
3. *Standard_dev* - the standard deviation of the distribution of X (for this task, enter **.3**)
4. *Cumulative* - Enter TRUE/FALSE (for this task as we require the cumulative probability, enter **TRUE**)

Then click **OK**.

Now you will see the value of P(X$\leq$3) in cell B1. To calculate P(X>3), in cell A2, type the label 'P(X>3)=' and in cell B2, type **=1-B1**.

Figure 7.2 NORMDIST Function dialogue box

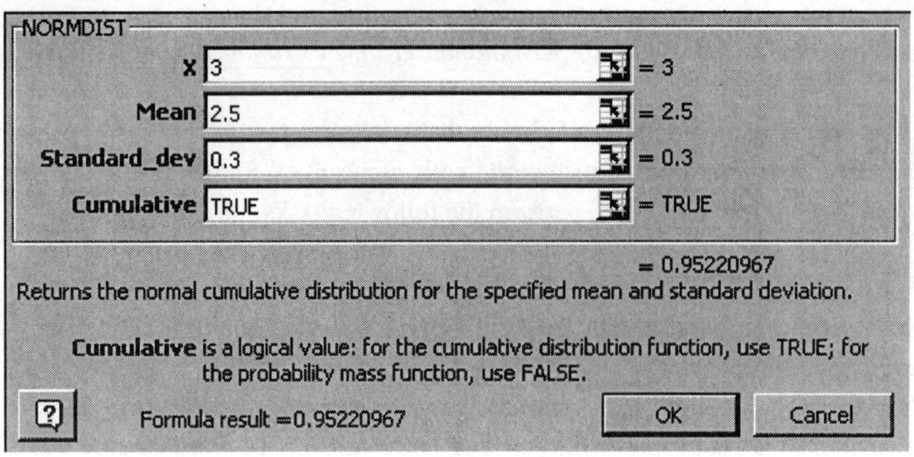

<u>*Alternatively*</u>, to obtain P(X≤ 3) in cell B1, simply type the formula **=NORMDIST(3,2.5,.3,TRUE)**; and to obtain P(X>3), in cell B2, type the formula **=1-NORMDIST(3,2.5,.3,TRUE)** or **=1-B1**.

Task 2 The expected level of sales is μ=2.5m. Therefore, the probability that the sales will fall within $.15m of the expected level of sales is P(2.5-.15≤X≤2.5+.15). In cell A3, type the label 'P(2.5-.15<=X<=2.5+.15)='. Note that P(2.5-.15≤X≤2.5+.15) = P(2.35≤X≤2.65) = P(X≤2.65) - P(X≤2.35). Therefore, to obtain P(2.5-.15≤X≤2.5+.15), in cell B3, type the formula **=NORMDIST(2.65,2.5,.3,TRUE)-NORMDIST(2.35,2.5,.3,TRUE)**.

Task 3 We need P(X>1.8). In cell A4, type the label 'P(X>1.8)='. Note that P(X>1.8)=1-P(X≤1.8). Therefore, to obtain P(X>1.8), in cell B4, type the formula **=1-NORMDIST(1.8,2.5,.3,TRUE)**.

Task 4 We need the sales value x_0 (say) such that the probability of X exceeding a value x_0 is equal to .09, i.e. P(X>x_0)=.09. Therefore, to obtain the value x_0 that gives a right tail value of .09, we need a cumulative probability of .91 i.e. P(X≤x_0)=1-.09=.91 (see diagram below).

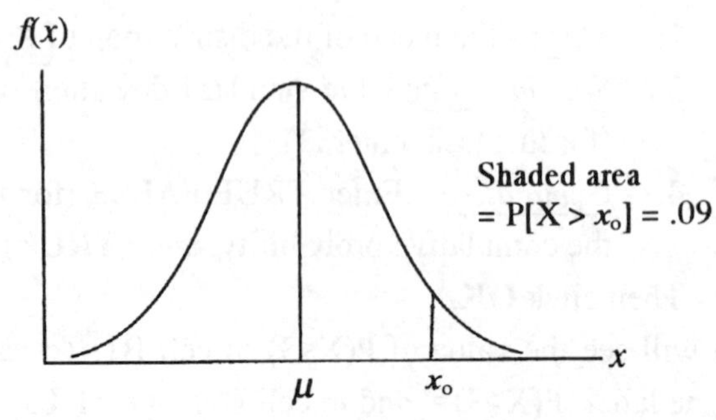

In cell A5, type the label 'x0='. The unknown x_0 can be obtained by *either* using the *Function Name* **NORMINV** in the *Function Category* **Statistical** or **All** in fx **Paste Function** by following a 2-step procedure as in NORMDIST (see Task 1) *or* by typing the formula (as discussed in Section 7.1), **=NORMINV(.91,2.5,.3)** in cell B5. The output for Tasks 1-4 is shown in the table below.

	A	B
1	P(X<=3)=	0.95221
2	P(X>3)=	0.04779
3	P(2.5-.15<X<2.5+.15)=	0.382925
4	P(X>1.8)=	0.990185
5	X0=	2.902226

Task 5 Save your EXCEL workbook as **A:EX71.XLS** on your disk and print the file.

7.4 Example 7.2: Graphing the normal distribution

The dividend yield of Australian firms in a particular year varied between 0 and 10 percent. The dividend yields follow approximately a normal distribution. Graphs of probability density functions are considered as useful tools to investment analysts in making decisions about where to invest their clients' money. Use EXCEL to perform the following tasks.

Task 1 Graph the probability density functions of the yields of three sectors of the firms under the following situation: Mean yields of all three sectors are the same and equal to 4 percent and standard deviations are 1, .5 and .25 percent, respectively. Manually, mark the mean value on your diagram.

Task 2 Graph the probability density functions of the yields of three sectors of the firms under the following situation: Standard deviations of the three sectors are the same and equal to 1 percent and the mean yields of the three industries are 5, 6 and 7 percent. Manually, mark the mean values on your diagram.

Task 3 Save your workbook as 'A:EX72.XLS' on your disk and print the file.

Solution to Example 7.2

As discussed in Section 7.1, if X ~ $\text{Normal}(\mu,\sigma^2)$ where μ is the mean and σ is the standard deviation of X, to obtain the probability density value $f(x)$ for a normal random variable X at x, we can use the EXCEL function NORMDIST(x,μ,σ,FALSE).

Task 1 ***Graphing the normal probability density functions with the same mean but different variances***

To obtain the probability density function values, first we divide the range of dividend yields (X) from 0 to 10 percent into very small intervals (for example, x=0, .1, .2, …., 9.9, 10) and enter them in column A. In cell A1, type 'X=x'. Click on cell A2 and type **0**, click on **Edit** on the *Menubar,* select **Fill** and then **Series…** on the attached dialogue box (see Figure 7.3).

Figure 7.3 Menubar and Edit menu

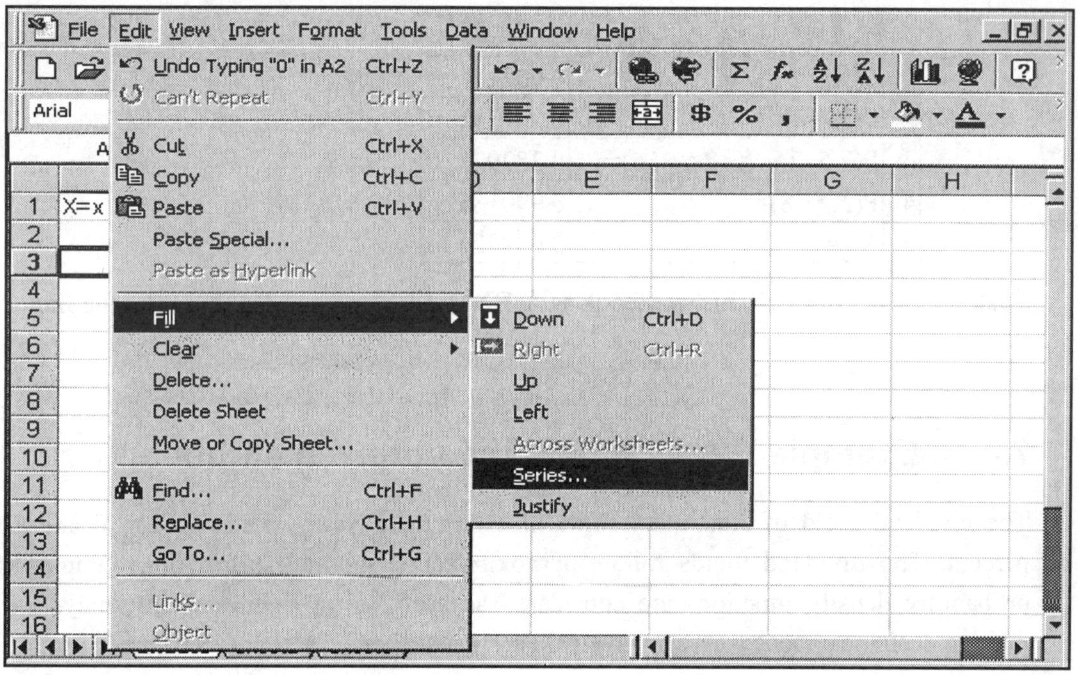

On the *Series dialogue box* appearing on the screen, select **Column** for *Series in*, **Linear** for *Type*, type **0.1** for the *Step value* and type **10** for *Stop value* and, then click **OK** (see Figure 7.4). [*Alternatively*, after typing **0** in cell A2 and **0.1** in cell A3, highlight cells A2 and A3, and copy onto cells A4 to A102 using the pointer +.]

Figure 7.4 Series Type dialogue box

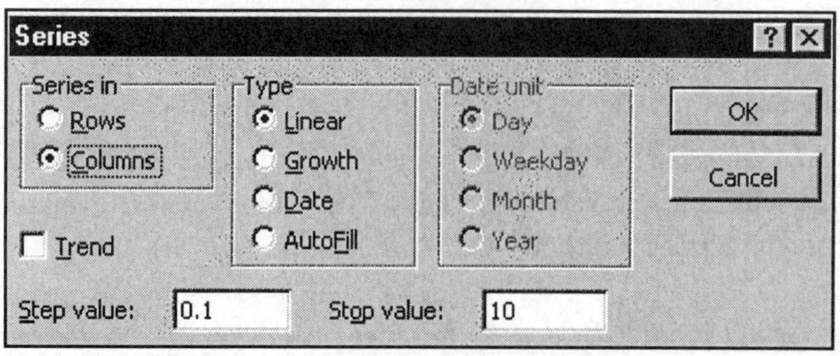

Now we will enter the mean value and the variances for the 3 sectors in column H. In cell H1, type the mean value **4** and in cells H2, H3 and H4, type the variances **1**, **.5** and **.25**, respectively. In cells B1, C1 and D1, type column headings for the three probability density functions, 'pdf1', 'pdf2' and 'pdf3'. In cell B2, type the formula =**NORMDIST(A2,$H$1,$H$2,FALSE)**. Then copy this formula onto cells B3 to B102. Now the probability density function $f(x)$ for X ~ Normal(4, 1^2) will appear in cells B2 to B102 of Column B. Similarly, you can obtain the f(x) values for X ~ Normal(4, $.5^2$) and X ~ Normal(4,$.25^2$). [Type the formulas =**NORMDIST(A2,$H$1,$H$3,FALSE)** in cell C2 and =**NORMDIST(A2,$H$1,$H$4,FALSE)** in cell D2, and then copy them onto cells C3 to C102 and D3 to D102, respectively.] Now highlight cells A1 to D102 and use **ChartWizard** to plot these three probability density functions. The resulting output is given in Figure 7.5.

Figure 7.5 Normal distributions with equal means and unequal variances

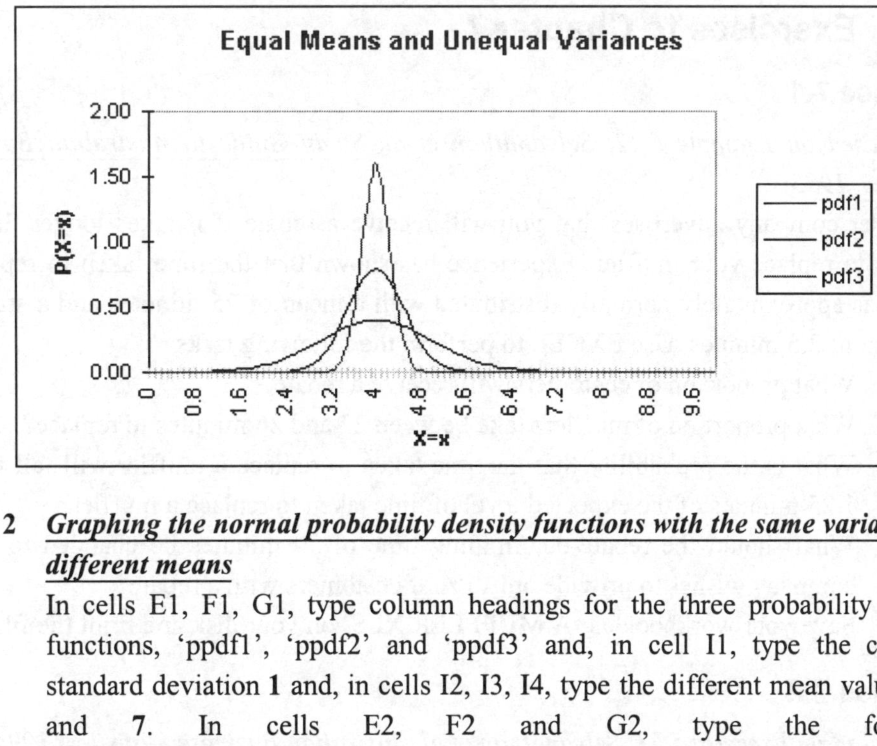

Task 2 *Graphing the normal probability density functions with the same variance but different means*

In cells E1, F1, G1, type column headings for the three probability density functions, 'ppdf1', 'ppdf2' and 'ppdf3' and, in cell I1, type the common standard deviation **1** and, in cells I2, I3, I4, type the different mean values **5**, **6** and **7**. In cells E2, F2 and G2, type the formulas =**NORMDIST(A2,$I$2,$I$1,FALSE)**, =**NORMDIST(A2,$I$3,$I$1,FALSE)** and =**NORMDIST(A2,$I$4,$I$1,FALSE)**, respectively, and copy them onto cells E3 to E102, F3 to F102 and G3 to G102, respectively. Now graph the three probability density function values in columns E, F and G using **ChartWizard** by typing the *Range* in Step 1 as **A1:A102,E1:G102**. The resulting output is given in Figure 7.6.

Task 3 Save your workbook as **A:EX72.XLS** on your disk and print the file.

Figure 7.6 Normal distributions with unequal means and equal variances

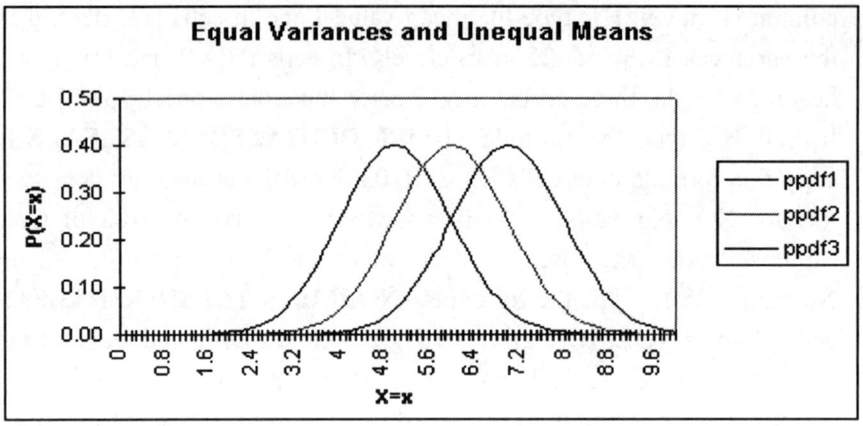

7.5 Exercises to Chapter 7

Exercise 7.1

(Ref: Based on Example 5.12, Selvanathan et al, Study Guide to Australian Business Statistics, 1995).

A muffler company advertises that you will receive a rebate if it takes longer than 30 minutes to replace your muffler. Experience has shown that the time taken to replace a muffler is approximately normally distributed with a mean of 25 minutes and a standard deviation of 2.5 minutes. Use EXCEL to perform the following tasks.

Task 1 What proportion of customers will receive a rebate?

Task 2 What proportion of mufflers take between 22 and 26 minutes to replace?

Task 3 What is the probability that the time taken to replace a muffler will fall within 1.25 minutes of the expected level of time taken to replace a muffler?

Task 4 What should the rebate-determining time of 30 minutes be changed to if the company wishes to provide only 1% of customers with a rebate?

Task 5 Save your workbook as 'A:MUFFLER.XLS' on your disk and print the file.

Exercise 7.2

(Ref: Based on Exercise 5.53, Selvanathan et al, Australian Business Statistics, 1994).

The lifetime of a certain brand of tyres is approximately normally distributed with a mean of 65,000 kilometers and a standard deviation of 2,500km. The tyres carry a warranty for 60,000km. Use EXCEL to perform the following tasks.

Task 1 What proportion of tyres will fail before the warranty expires?

Task 2 What proportion of tyres will fail after the warranty expires?

Task 3 What proportion of tyres will fail after the warranty expires but before they have lasted for 61,000 kilometers?

Task 4 Save your EXCEL workbook as 'A:TYRE.XLS' on your disk and print the file.

Chapter 8

Sampling and Sampling Distributions

Populations are generally quite large. In most cases, it is not possible to collect data for the whole population. Therefore, to make inferences about a population, rather than investigating the entire population, we rely on random samples drawn from the population. Our first task is to draw a representative random sample of the population. Based on the nature of the statistical problem to be solved, various sampling plans (methods) are available to select a sample, for example, simple random sampling, stratified random sampling, cluster sampling, purposive sampling etc. The examples given in this chapter uses the most basic sampling plan, the *Simple Random Sampling* (with or without replacement). When simple random sampling is done with replacement, each member of the population is equally likely to be included in the sample.

8.1 Sampling distribution of the sample mean and its mean and variance

If we repeatedly draw random samples from a population, for each sample, the value of the sample mean ($\overline{X}$) will vary. Therefore, sample mean is a random variable which obviously has a distribution, called the *sampling distribution of the sample mean.* The sampling distribution of $\overline{X}$ allows us to make probability statements about the sample mean based on the values of the sample size *n* and the population parameters μ and σ^2. If the population is normally distributed, then the sampling distribution of the sample mean is also normal.

The mean of $\overline{X}$ is equal to the population mean μ (mean of the population random variable X) and the variance of $\overline{X}$ is equal to the population variance σ^2 divided by the sample size *n*. That is,

$$\mu_{\overline{X}} = \mu$$

and

$$\sigma_{\bar{x}}^2 = \frac{\sigma^2}{n}$$

In most problems we deal with, the population will be very large. However, whenever the population size is small, it becomes necessary to multiply the variance of the sample mean by a *Finite Population Correction Factor* (FPCF). If N is the population size and n is the sample size, then if $n/N > .05$ (i.e., the sample size is relatively large compared to the population size),

$$\sigma_{\bar{x}}^2 = \left(\frac{N-n}{N-1}\right)\frac{\sigma^2}{n}$$

Aim: At the completion of the examples in this chapter you will be able to use **EXCEL** to select random samples (with or without replacement) from a population, generate random numbers from a normal and non-normal distribution, calculate probabilities and verify the Central Limit Theorem.

8.2 Example 8.1: Sampling distribution of the sample mean when samples are selected from a normal population

A marketing manager of a leading firm believes that total sales for his firm follows a normal distribution with mean \$2.5million and standard deviation \$.3 million.

Task 1 Generate 1000 random numbers from its total sales population [distributed Normal(2.5,.3^2)] for this firm and set them up in a 100 rows by 10 columns table. Draw the histogram for the 1000 random sales figures. Calculate the mean and standard deviation of the 1000 random sales figures.

Task 2 Consider each row as a sample of size 10 so that you have 100 samples from a normal population with mean = 2.5 and standard deviation = .3. Compute the mean value for each sample (row of 10 values). Draw the histogram for the 100 sample means. Calculate the mean and standard deviation of the 100 sample means.

Task 3 Comment on the shape of the two histograms and the two sets of mean and standard deviation values obtained in Tasks 1 and 2.

Task 4 Save the workbook as 'A:EX81.XLS' on your disk and print the file.

Solution to Example 8.1

Let X be the total sales. The random variable X has a normal distribution with mean 2.5 and standard deviation .3.

Task 1 ***Generating random numbers from a normal population with a given mean and standard deviation***

To generate random numbers, first click the **Tools** option on the *Menubar* and select the **Data Analysis**… option. From the Data Analysis menu now select **Random Number Generation** and click **OK**. The *Random Number Generation dialogue box* will appear on the screen (see Figure 8.1). To generate 1000 random sales figures of the firm and to set them up in a 100 by 10 table, we enter **10** for the *Number of Variables* and **100** for the *Number of Random Numbers* in the dialogue box.

Figure 8.1 Random Number Generation dialogue box

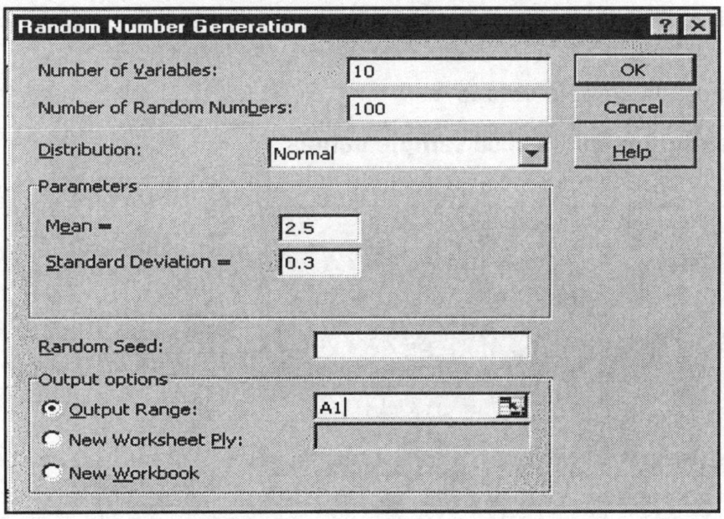

In the *Distribution* cell use the down arrow to select **Normal** distribution and enter in the parameters box *Mean* = **2.5** and *Standard Deviation* = **.3**. Leave *Random Seed* blank. Under *Output options* click *Output Range:* and in the blank space next to it enter **A1** and then click **OK**. This will generate 1000 random numbers from a normal population with mean 2.5 and variance .3 and put them in a 100 rows by 10 columns table.

Note that, as the random numbers are generated randomly, unless a constant Random Seed is assigned, one will generate different sets of numbers each time.

Calculating the mean and standard deviation of the 1000 sample values and constructing their distribution

To obtain the mean and standard deviation of the 1000 sample values, in cell A101 and A102, type the labels 'Mean =' and 'SD='. In cells B101 and B102, type the formulas **=AVERAGE(A1:J100)** and **=STDEV(A1:J100)**, respectively. To draw the histogram of the 1000 sample values, we use the Histogram option under the Data Analysis tool. First click the *Tools* option on the *Menubar* and select **Data Analysis**… From the *Data Analysis* menu, select

Histogram (see Figures 3.1-3.3) and click **OK**. Now on the *Histogram dialogue box*, under Input options, enter the *Input Range* as **A1:J100**. Leave *Bin Range* and *Labels* blank and, under Output options, tick the square for *Chart Output* and enter *Output Range* as **A120**. Then click **OK**. This will give the histogram for the distribution of the 1000 sample values similar to the one shown in Figure 8.2. [As the bars are disjoint, you have to edit the chart as shown under *Removing the Gaps between the Bars of a Histogram Chart (Editing a Chart)* in Section 3.4. Also, the X-axis labels can be made to one decimal as in Figure 8.2 by reducing the number of decimals of the entries in the *Bin* column.] Since the population is normally distributed, as one would expect, the histogram of these 1000 sample values is also symmetric and bell-shaped (see Figure 8.2).

Figure 8.2 Distribution of the sample values

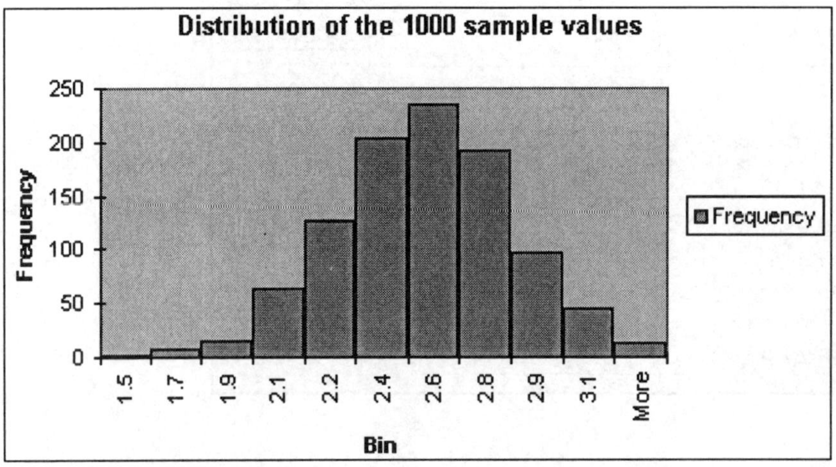

Task 2 ***Calculating the mean and standard deviation of the sample means and constructing the sampling distribution of the sample mean***

Considering each row as a sample of size 10, we have 100 samples. To obtain the mean value of the first sample of size 10, i.e. the 10 sales values in the first row, in cell K1, type the formula **=AVERAGE(A1:J1)**. To obtain the averages of the remaining rows, copy the formula in cell K1 into cells K2 to K100 using the pointer +. Use the Data Analysis tool, *Histogram,* to draw the histogram for these 100 mean values (see Figure 8.3). In cells A103 and A104, enter the labels 'Mean of Xbar=' and 'SD of Xbar ='. In cells B103 and B104, type the formulas **=AVERAGE(K1:K100)** and **=STDEV(K1:K100)**, respectively.

Task 3 ***Comparing the distribution of the sample values and the sample mean***

Looking at the sampling distribution of the sample mean in Figure 8.3 (the histogram of the 100 sample means), we see that it is bell-shaped and symmetric as the distribution of the population, which is normal. That is, the sample mean is also normally distributed. As the population is normal, the distribution of the 1000 sample values is also normal (see Figure 8.2).

Figure 8.3 Sampling distribution of the sample mean

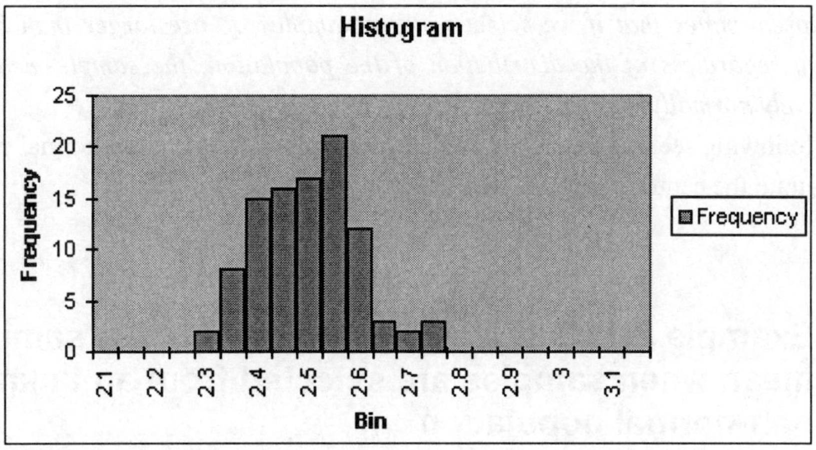

Comparing the means and the standard deviations

Comparing the mean value of the sample mean in cell B103 (=2.473) with the population mean of 2.5, we can verify that the mean of the sample means $= \mu_{\bar{x}} = \mu =$ population mean. To calculate the standard error, in cell A105, type the label 'Sigma/SQRT(n)=' and, in cell B105, type the formula **=.3/SQRT(10)**. The mean and standard deviation output in Tasks 1-3 are given below.

	A	B
101	Mean=	2.473
102	SD=	0.294
103	Mean of Xbar=	2.5
104	SD of Xbar=	0.096
105	Sigma/SQRT(n)=	0.095

Comparing the value in cell B104 (which is $\sigma_{\bar{x}}$=0.096) with the one in cell B105 (which is $\sigma/\sqrt{n}$=0.095), we can see that they are very close to each other, which verifies that the standard error of the sample mean is equal to the standard deviation of the population divided by the square root of the sample size, $\sigma_{\bar{x}} = \sigma/\sqrt{n}$.

Task 4 Save your workbook as **A:EX81.XLS** on your disk and print the file.

8.3 Central Limit Theorem

In Example 8.1, we generated random samples from a normal population with a given mean and standard deviation. We also demonstrated that the sampling distribution of the sample mean $\overline{X}$ is also normally distributed. In the next two sections, we demonstrate that even if the population is not normal, the sampling distribution of the sample mean $\overline{X}$ will be approximately normal provided the sample size is reasonably large (i.e., $n \geq 30$).

This important result is known as the **Central Limit Theorem**. *In a nut shell, Central Limit Theorem states that if we select random samples of size larger than 30 from a population, regardless of the distribution of the population, the sample mean will be approximately normally distributed.*

In the following section we draw samples from some unknown non-normal population to demonstrate the central limit theorem.

8.4 Example 8.2: Sampling distribution of the sample mean when samples are selected from an unknown non-normal population

A department-store audit involves checking a random sample from a population of 100 outstanding credit-account balances. The 100 accounts are listed in the table below.

1	25	**26**	291	**51**	35	**76**	876
2	0	**27**	16	**52**	67	**77**	561
3	605	**28**	0	**53**	432	**78**	348
4	1010	**29**	402	**54**	211	**79**	900
5	527	**30**	31	**55**	62	**80**	1020
6	34	**31**	245	**56**	34	**81**	123
7	245	**32**	342	**57**	57	**82**	23
8	59	**33**	34	**58**	89	**83**	4
9	67	**34**	46	**59**	76	**84**	345
10	403	**35**	76	**60**	49	**85**	456
11	918	**36**	34	**61**	234	**86**	321
12	801	**37**	475	**62**	341	**87**	531
13	227	**38**	234	**63**	23	**88**	234
14	0	**39**	345	**64**	0	**89**	567
15	47	**40**	675	**65**	890	**90**	943
16	0	**41**	346	**66**	456	**91**	289
17	102	**42**	456	**67**	345	**92**	421
18	215	**43**	65	**68**	390	**93**	23
19	429	**44**	323	**69**	453	**94**	0
20	197	**45**	675	**70**	564	**95**	34
21	159	**46**	43	**71**	563	**96**	23
22	279	**47**	12	**72**	87	**97**	90
23	115	**48**	0	**73**	56	**98**	237
24	27	**49**	34	**74**	28	**99**	419
25	27	**50**	2	**75**	678	**100**	88

Use EXCEL to perform the following tasks.

Task 1 Draw a histogram of the population of 100 accounts. Comment on the shape of the distribution. Is the distribution approximately a bell-shaped? Determine the population mean and standard deviation.

Task 2 Select 25 samples from this population each of them containing 30 accounts selected at random with replacement. For each sample, calculate the mean value. Calculate the mean and standard error of the sample mean of these 25 sample mean values. Compare the mean and standard error of the sample mean with the population mean and standard deviation.

Task 3 Draw a histogram of the 25 sample mean values. Comment on the shape of the sampling distribution of the sample mean. Is the distribution approximately bell-shaped? Verify the Central Limit Theorem from your results.

Task 4 Save the workbook as 'A:EX82.XLS' on your disk and print the file.

Solution to Exercise 8.2

Task 1 First, type the column headings 'ACCOUNT NUMBER' and 'AMOUNT' in cells A1 and B1, respectively. Then click on cell A2 and type **1**. Click on **Edit** on the *Menubar*, select **Fill** and then **Series...** from the dialogue box. On the *Series dialogue box* appearing on the screen, select **Column** for *Series in*, **Linear** for *Type*, type **1** for the *Step value* and type **100** for *Stop value* and then click **OK**. [*Alternatively*, after typing **1** in cell A2, type formula **=A2+1** in cell A3. Then copy it onto cells A3 to A101 using the pointer+. Otherwise, type **1** and **2** in cells A2 and A3, highlight the two cells and drag down up to cell A101.] Enter the data in cells B2 to B101.

Graphing the distribution of the population

To draw the histogram of the 100 population values entered in Column B, we use the Histogram option under the Data Analysis tool. First click the *Tools* option on the *Menubar* and select **Data Analysis**... From the *Data Analysis* menu, now select **Histogram** (see Figures 3.1-3.3) and click **OK**. Now on the Histogram dialogue box, under Input options, enter the *Input Range* as **B2:B101**. Leave *Bin Range* and *Labels* blank and, under Output options, tick the square box against *Chart Output* and enter *Output Range* as **C40**. Then click **OK**. This will give the histogram for the distribution of the population (see Figure 8.4). [As the bars are disjoint, you have to edit the chart as shown under *Removing the Gaps between the Bars of a Histogram Chart (Editing a Chart)* in Section 3.4.] As can be seen, the histogram shows that the distribution of the population is skewed to the right. Thus the distribution of the population is not symmetric and does not have a bell shape.

To obtain the mean and standard deviation of the population in cells A102 and A103, type the labels 'Pop Mean=' and 'Pop STD='. In cells B102 and B103, respectively, type the formulas **=AVERAGE(B2:B101)** and **=STDEVP(B2:B101)**.

Figure 8.4 Distribution of the population

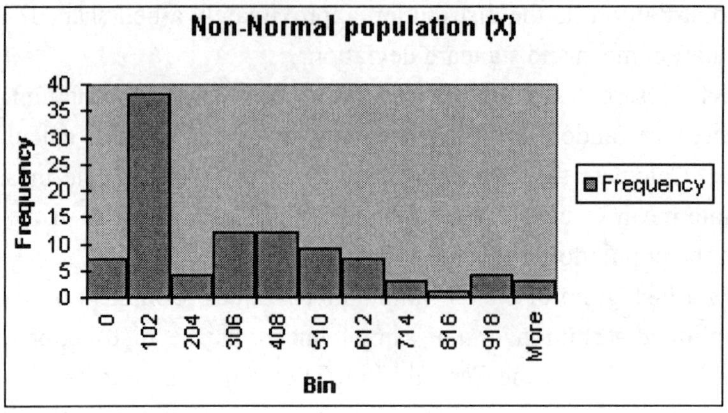

Task 2 ***Selecting random samples of a given size***

To select a random sample of a given size (here, n=30) from a population we use the Sampling option under the Data Analysis tool. In cell C1, type label 'Sample 1'. First click the **Tools** option on the *Menubar* and select the **Data Analysis…** option. From the *Data Analysis menu* now select **Sampling** and click OK. Now on the sampling dialogue box under Input options enter the *Input Range* as **B2:B101**, select *Sampling Method* as **Random** and type *Number of samples*: as **30**. [Note that, number of samples here means the sample size.] Under Output options, click the circle for Output range and in the blank space next to it, enter Output range as C2. Then Click **OK**. This will generate a random sample of size 30 in column C (in cells C2 to C31) from the given population in column B.

In cells, D1 to AA1, type labels 'Sample 2', 'Sample 3',…. 'Sample 25'. [*Alternatively*, copy the content in cell C1 into cells D1 to AA1 by using the pointer ✛]. We repeat the sampling selection to obtain 25 such samples by entering output ranges as **D2, E2, ….Z2** and **AA2** with all other entries remaining the same as before. Part of your output will be similar Figure 8.5. [Due to 'random' generation, your numbers will not match those in Figure 8.5.]

Figure 8.5 Part of the generated random samples

	A	B	C	D	E	F	G	H
1	ACCOUNT NUMBER	AMOUNT	Sample 1	Sample 2	Sample 3	Sample 4	Sample 5	Sample 6
2	1	25	28	348	159	89	65	531
3	2	0	34	348	65	605	88	348
4	3	605	0	12	876	943	0	31
5	4	1010	323	2	23	291	23	346
6	5	527	563	1010	67	57	35	234
7	6	34	46	31	527	531	943	34
8	7	245	237	323	456	900	323	456
9	8	59	237	1010	31	900	43	34
10	9	67	341	56	890	23	159	90
11	10	403	46	34	876	419	90	323

Calculating the mean and standard error of the sample mean

To obtain the mean of the sample in column C, in cell C32, type the formula **=AVERAGE(C2:C31)**. We then copy the formula in cell C32 using the pointer + onto cells D32 to AA32, to obtain the mean of the remaining 24 samples. To calculate the mean and standard error of these 25 sample means in cell A104 and A105, type the labels 'Mean(XBAR)=' and 'SD(XBAR)='. In cell B104 and B105, enter the formulas **=AVERAGE(C32:AA32)** and **=STDEV(C32:AA32)** for the mean and standard error of the sample mean. As can be seen, the mean of the sample means (can be considered an approximate value for the expected value of the sample mean, E[$\overline{X}$]) in cell B104 is very close to the population mean in cell B102. To verify that the standard error of the sample mean is approximately equal to the population standard deviation divided by the square root of sample size, in cell C105, type the formula **=B103/SQRT(30)**. Comparing the values in cells B105 and C105, we conclude that standard error of the sample mean is approximately equal to the population standard deviation divided by the square root of the sample size (n=30).

Task 3 ***Graphing the sampling distribution of the sample mean***

As in Task 1, draw the histogram for the 25 sample mean values, by typing the *Input Range* as **C32:AA32** and *Output Range* as **C70**. The output will be similar to Figure 8.6. As can be seen, even though the population histogram in Figure 8.4 was not symmetric or bell-shaped (i.e., non-normal) the histogram of the sample means in Figure 8.6 (which is the sampling distribution of the sample mean) appears to be bell-shaped (i.e., normal).

Figure 8.6 Sampling distribution of the sample mean

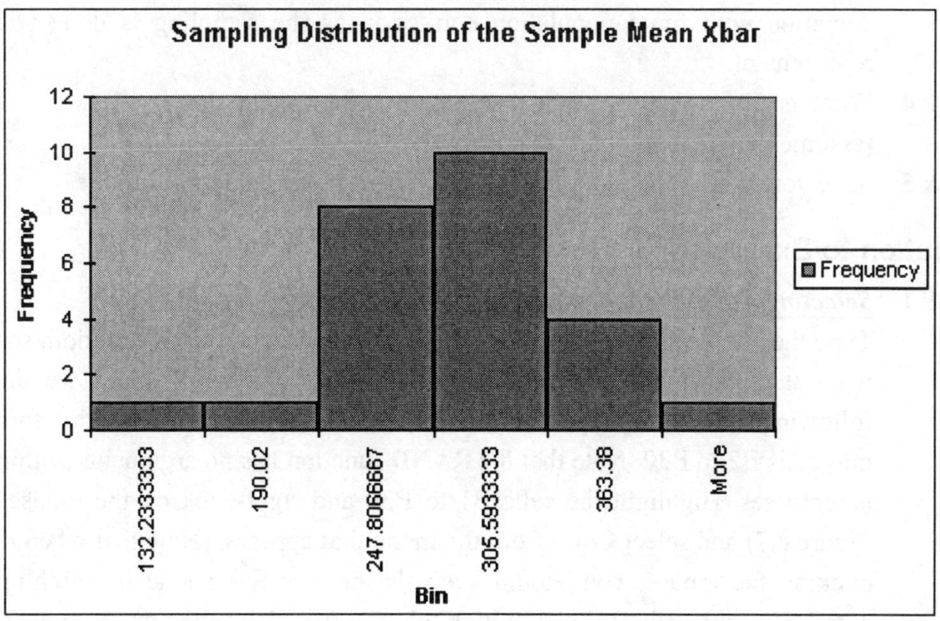

From the above results we can conclude that when the sample size is greater than or equal to 30, the distribution of the sample mean $\overline{X}$ is approximately normal even if the population from which the samples are drawn is not normal. This verifies the central limit theorem.

Task 4 Save the workbook as **A:EX82.XLS** on your disk and print the file.

8.5 Example 8.3: Sampling without replacement

In Example 8.2, we generated simple random samples with replacement. Now in the following example, we generate simple random samples without replacement.

A weight reduction programme advertised that they have developed a new diet method to reduce weights by at least 20kg in twenty weeks. Their advertisement offered a free introductory offer for the first 30 people in the weight range of 80 to 100 kg. The following are the weight losses (in kg) of the 30 people who were selected for the free introductory programme after 20 weeks.

10.5 1.3 4.7 20.7 10.7 18.7 23.5 5.6 14.5 22.5 10.9 24.5 9.8 12.7 7.6
6.4 18.5 17.4 20.0 19.8 17.6 12.5 3.7 24.9 22.6 24.7 20.1 19.8 20.4 6.8

Task 1 Select a random sample of 10 weight losses without replacement from the above 30 weight losses (population). Calculate the mean and standard deviation of your sample.

Task 2 Calculate the population (30 weight losses) mean and standard deviation.

Task 3 Calculate the standard error of the sample mean from the population standard deviation with finite population correction as the sampling is done without replacement.

Task 4 What is the probability that the sample mean weight loss will exceed 20kg (assume that the population is normal)?

Task 5 Save your workbook as 'A:EX83.XLS' on your disk and print the file.

Solution to Example 8.3

Task 1 ***Selecting a simple random sample without replacement***

Type the 30 weight losses in cells A1 to A30. To select a simple random sample without replacement of size 10 from this 30 population values we do the following. In cell B1, type the formula =**RAND( )** and then copy this formula into cells B2 to B30. Note that the **RAND** function has no arguments within the parentheses. Highlight the cells B1 to B30 and right-click on the mouse (see Figure 8.7) and select **Copy** from the menu that appears. [Note that when right-clicking the mouse, you should keep the pointer ✚ inside the highlighted area.] With the cells B1 to B30 highlighted again right-click on the mouse and now select **Paste Special** (see Figure 8.8).

Figure 8.7 Simple random number generation - RAND() Copy

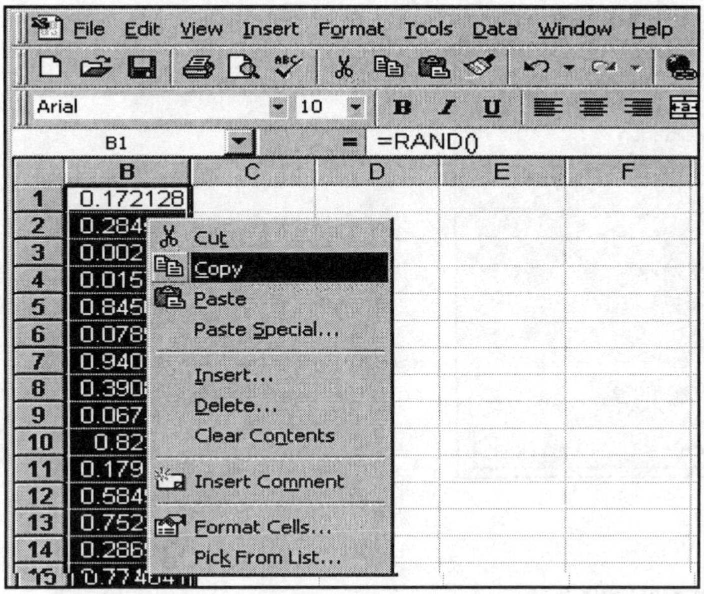

Figure 8.8 Simple random number generation - RAND() Paste Special

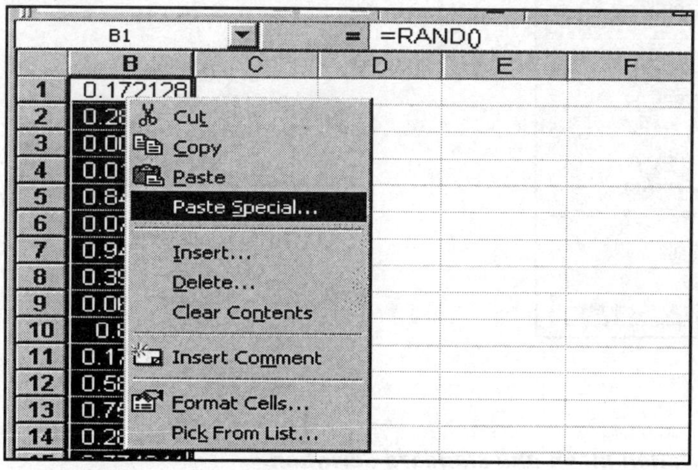

In the *Paste Special* dialogue box (see Figure 8.9), tick on **Values** under *Paste* and **None** under *Operation* and leave the boxes for **Skip Blanks** and **Transpose** as blank and click **OK**.

Now highlight all 30 entries in columns A and B and select **Data** on the *Menubar* and then select **Sort….** Select *Sort by* **Column B** (see Figure 8.10) and click on **Ascending** and then click **OK**. Now the first 10 values in cells A1 to A10 will form a simple random sample without replacement.

To obtain the mean of this sample in cells C1 and C2, type the labels 'Sample mean=' and 'Sample SD=' and in cells D1 and D2, type the formulas **=AVERAGE(A1:A10)** and **=STDEV(A1:A10)**, respectively.

Figure 8.9 Paste Special dialogue box

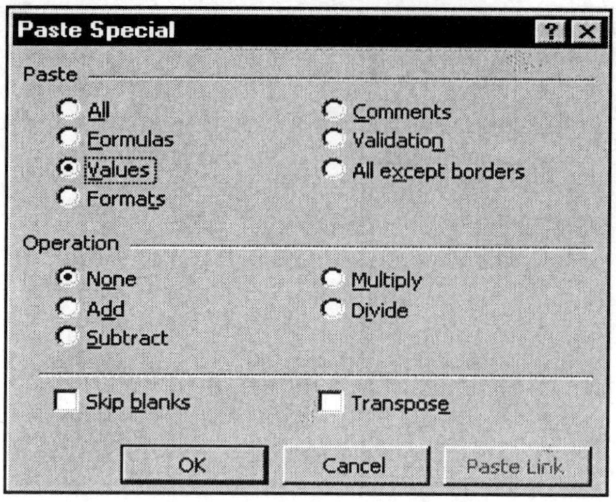

Figure 8.10 SORT Function dialogue box

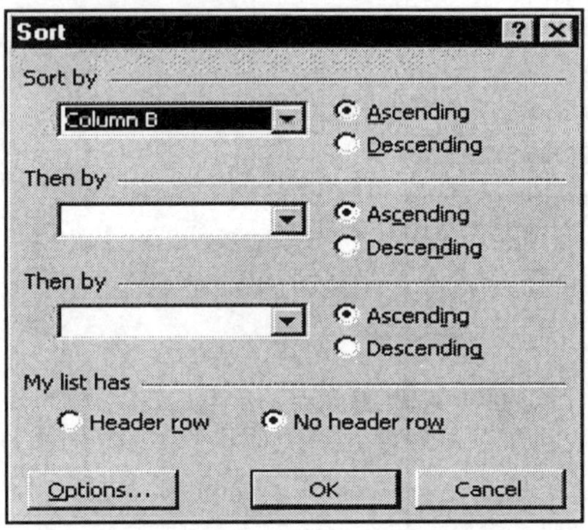

Task 2 ***Calculating the population mean and standard deviation***

In cells C4 and C5, type the labels 'Population mean=' and 'Population SD='. In cells D4 and D5, type the formulas =**AVERAGE(A1:A30)** and =**STDEVP(A1:A30)**.

Task 3 ***Calculating the standard error of the sample mean with finite population correction***

As the population size $N = 30$ and sample size $n = 10$, $n/N = 10/30 = .33>.05$. That is, the sample is not a relatively small sample. Therefore, when sampling is done without replacement, to calculate the standard error of the sample mean, we need to multiply it by the square root of *the Finite Population Correction Factor* (FPCF), $(N\text{-}n)/(N\text{-}1)$. In cell C6, type the label 'SE of Xbar with FPCF=' and, in cell D6, type the formula =**(D5/SQRT(10))*SQRT((30-10)/(30-1))**.

Task 4 In cell C8, type the label 'P(Xbar>20)='. First, we calculate the standardised value of 20. In cell C7, type 'Z-value=' and, in cell D7, type the formula **=(20-D4)/D6**. In cell D8, type the formula **=1-NORMSDIST(D7)**. *Alternatively*, without calculating the standardised value, we can straight-away type the formula **=1-NORMDIST(20,D4,D6,TRUE),** where D4 and D6 are the cell locations of the mean and standard error (with FPCF) of $\overline{X}$. Figure 8.11 gives the output for Tasks 1-4.

Figure 8.11 Standard error of the sample mean and probability

	C	D
1	Sample mean=	11.91
2	Sample SD=	5.85898
3		
4	Population mean=	15.1133
5	Population SD=	6.9895
6	SE of Xbar with FPCF=	1.83553
7	Z-value=	2.66226
8	P(Xbar>20)=	0.00388

Task 5 Save your workbook as **A:EX83.XLS** on your disk and print the file.

8.6 Exercises to Chapter 8

Exercise 8.1

A supermarket in Freemantle receives cereal boxes from a company in Melbourne. The manager is told that the weights of the cereal boxes are normally distributed with a mean of 165 grams and a standard deviation of 5 grams.

Task 1 Generate 1000 random weight figures of the cereal boxes from the population and set them up in a 100 rows by 10 columns table. Draw a histogram for the weights of the 1000 random cereal boxes. Calculate the sample mean and standard deviation of the 1000 random cereal box weights.

Task 2 Consider each row as a sample of size 10 so that you have 100 samples from a normal population with mean = 165 and standard deviation = 5. Compute the mean value for each sample. Draw the histogram for the 100 sample means. Calculate the mean and standard deviation of the 100 sample means.

Task 3 Comment on the shape of the two histograms and the two sets of mean and standard deviation values obtained in Tasks 1 and 2.

Task 4 What is the probability that the weight of a cereal box will exceed 210 grams?

Task 5 What is the probability that the weight of a cereal box will fall within 2 grams of the expected weight?

Task 6 Determine the weight level that has only a 10 percent chance of being exceeded.

Task 7 Save the workbook as 'A:CEREAL.XLS' on your disk and print the file.

Exercise 8.2

The following are the heights (in cms) of the 100 students attending the Pioneer Primary School in Queensland.

1	95	**26**	110	**51**	106	**76**	118		
2	160	**27**	124	**52**	134	**77**	123		
3	120	**28**	96	**53**	141	**78**	135		
4	92	**29**	108	**54**	127	**79**	147		
5	148	**30**	143	**55**	135	**80**	156		
6	165	**31**	134	**56**	161	**81**	163		
7	106	**32**	141	**57**	154	**82**	109		
8	109	**33**	118	**58**	148	**83**	110		
9	141	**34**	116	**59**	152	**84**	99		
10	135	**35**	137	**60**	98	**85**	126		
11	112	**36**	141	**61**	111	**86**	132		
12	146	**37**	108	**62**	124	**87**	143		
13	99	**38**	160	**63**	143	**88**	157		
14	105	**39**	143	**64**	135	**89**	162		
15	121	**40**	118	**65**	117	**90**	151		
16	109	**41**	109	**66**	126	**91**	149		
17	113	**42**	115	**67**	122	**92**	152		
18	100	**43**	110	**68**	161	**93**	116		
19	127	**44**	140	**69**	128	**94**	160		
20	141	**45**	117	**70**	139	**95**	128		
21	126	**46**	126	**71**	117	**96**	109		
22	135	**47**	132	**72**	131	**97**	124		
23	102	**48**	141	**73**	147	**98**	133		
24	131	**49**	109	**74**	153	**99**	112		
25	147	**50**	110	**75**	116	**100**	96		

Task 1 Draw a histogram of the population of 100 heights. Comment on the shape of the distribution. Is the distribution approximately bell-shaped? Determine the population mean and standard deviation.

Task 2 Select 25 samples from this population each containing 30 heights selected at random with replacement. For each sample, calculate the mean value. Calculate the mean and standard error of the sample mean of these 25 sample mean values. Compare the mean and standard error of the sample mean with the population mean and standard deviation.

Task 3 Draw a histogram of the 25 sample mean values. Comment on the shape of the sampling distribution of the sample mean. Is the distribution approximately bell-shaped? Verify the Central Limit Theorem from your results.

Task 4 Save the workbook as 'A:HEIGHT.XLS' on your disk and print the file.

Chapter 9

Interval Estimation

One way of making inferences about a population is by estimating the unknown population parameter. If the data type is quantitative, we will be estimating the population mean μ and if the data type is qualitative, we will be estimating the population proportion p. There are two types of estimators, namely, point estimator and interval estimator. A *point estimator* estimates the value of an unknown parameter, using a single value. An *interval estimator* estimates the value of an unknown parameter, using an interval which has a certain probability of including the value of the population parameter.

In general, the confidence interval for a population parameter is defined as

$$\text{Point estimate} \pm \text{B},$$

where B = interval coefficient × standard error of the point estimate. The interval coefficient is calculated based on the confidence level ($1 - \alpha$), which is the probability that the population parameter will fall inside the constructed interval. For example, when estimating the population mean μ of a normal population with known variance σ^2, the interval coefficient is equal to the z-value, $z_{\alpha/2}$. The value 2B is the width of the confidence interval. The end points of the confidence interval are called the *confidence limits*. The upper limit (Point estimate + B) is called the *upper confidence limit* (UCL) and the lower limit (Point estimate - B) is called the *lower confidence limit* (LCL).

Aim: At the completion of the examples in this chapter you will be able to use **EXCEL** to construct confidence intervals for the population mean, population proportion and to determine the required sample size to achieve a desired width for a confidence interval.

9.1 Constructing confidence intervals involving a single population: Quantitative data type

When the data type is quantitative, the problem objective is to estimate the population mean μ. We use the sample mean $\overline{X}$ as the point estimator for the population mean μ. [$\overline{X}$ is an unbiased estimator for μ.]

Case 1 ***When the population variance σ^2 is known***

The (1-α)100% confidence interval for μ is given by

$$\overline{X} \pm \text{B}, \quad \text{where } B = z_{\alpha/2}\frac{\sigma}{\sqrt{n}}$$

provided $\overline{X}$ is normally distributed, where the probability (1-α) is called the *level of confidence*. Note that $\overline{X}$ is normally distributed if either the population is normally distributed or if the distribution of the population is unknown or non-normal and the sample size $n \geq 30$ (using the Central Limit theorem).

Case 2 ***When the population variance σ^2 is unknown***

When the population variance is unknown, we replace σ with (its unbiased estimator) s and the standardised value

$$t = \frac{\overline{X} - \mu}{s/\sqrt{n}}$$

has a t-distribution with n-1 degrees of freedom (d.f.).

Therefore, the (1 - α)100% confidence interval for μ is given by

$$\overline{X} \pm \text{B}, \quad \text{where } B = t_{\alpha/2,n-1}\frac{s}{\sqrt{n}}$$

provided the population is normally distributed.

Case 3 ***When the population variance σ^2 is unknown and the sample size $n > 30$***

The t-distribution is a symmetric distribution with mean 0 and variance (n-2)/(n-1). As the sample size increases, the variance of the t-distribution approaches 1. Therefore, for large sample sizes (i.e., d.f. ≥ 30 or $n > 30$), we can approximate the t-distribution with the standard normal (z-) distribution (which is symmetric, bell-shaped, and has mean 0 and variance 1). Therefore, in this case, we can approximate the interval coefficient $t_{\alpha/2,n-1}$ (in Case 2) by $z_{\alpha/2}$ and the confidence interval can be written as

$$\overline{X} \pm \text{B}, \quad \text{where } B = t_{\alpha/2,n-1}\frac{s}{\sqrt{n}} \approx z_{\alpha/2}\frac{s}{\sqrt{n}}$$

The usefulness of a confidence interval estimator depends on the width or the precision of the interval. The smaller the interval the better the precision of the estimate. To control the width of a confidence interval, we select the optimum sample size for a given level of confidence. To obtain the required sample size, assuming σ is known, we use the following formula

$$n = \left(\frac{z_{\alpha/2}\sigma}{B}\right)^2$$

9.2 Example 9.1: Constructing confidence intervals for the population mean (μ) when σ is known

To determine the mean waiting time for his customers, a bank manager took a random sample of 40 customers and found that the mean waiting time was 7.2 minutes.

Task 1 Assuming that the population standard deviation is known to be 5 minutes, find the 95% confidence interval estimate of the mean waiting time for all of the bank customers.

Task 2 Save the workbook as 'A:EX91.XLS' on your disk and print the file.

Solution to Example 9.1

Let X be the waiting time for customers at the bank and, μ be the mean waiting time and σ be the standard deviation of the population.

Task 1 In cell A1, type the heading '95% Confidence Interval for the Single Population Mean Mu' and centre across A1 to G1. Type heading 'Population Standard Deviation Sigma is Known' in cell A2 and center it across A2 to G2. The sampling distribution of $\overline{X}$ will be distributed exactly as normal if the population is normally distributed or $\overline{X}$ will be distributed approximately as normal if $n \geq 30$ (based on central limit theorem). As the sample size in this case is n=40, $\overline{X} \sim$ approximate Normal. Therefore, when the population standard deviation σ is *known*, to construct a 95% confidence interval for the single population mean μ, we use the formula (Case 1, Section 9.1)

$$\overline{X} \pm \text{B, where B} = z_{.025}\frac{\sigma}{\sqrt{n}}$$

First, in cells A4, A5, A6 and A7, type the labels 'Sample size n =', 'Confidence level =' 'Sample mean Xbar =', and 'Population SD sigma =' to input the given information. In cells B4 to B7, type the corresponding values **40**, **.95**, **7.2** and **5**.

Calculating half the width (B) of the confidence interval

Now type 'Half width B =' in cell A9 and click on cell B9.

Method 1 To calculate B click on the icon *fx* **Paste Function** on the *Standard toolbar* and select the *Function Category* **Statistical** and *Function Name* **CONFIDENCE** and then click **OK**. The dialogue box that appears on the screen will ask for 3 values (see Figure 9.1): *Alpha, Standard_dev* and *Size*. For a 95% confidence interval, we input alpha as **.05** or **1-B5**. [Note that for a 95% confidence interval, the input for ALPHA (α) is .05 and not $\alpha/2$ = 0.25; and for a 90% confidence interval, we input the alpha value as .10.] For this task, Standard deviation is given as **5** or **B7** and the sample size is **40** or **B4**. So, input these 3 values and then click **OK**. EXCEL will give the output value for the half width of the confidence interval B and this value will now be stored in cell B9.

Figure 9.1 CONFIDENCE Function dialogue box

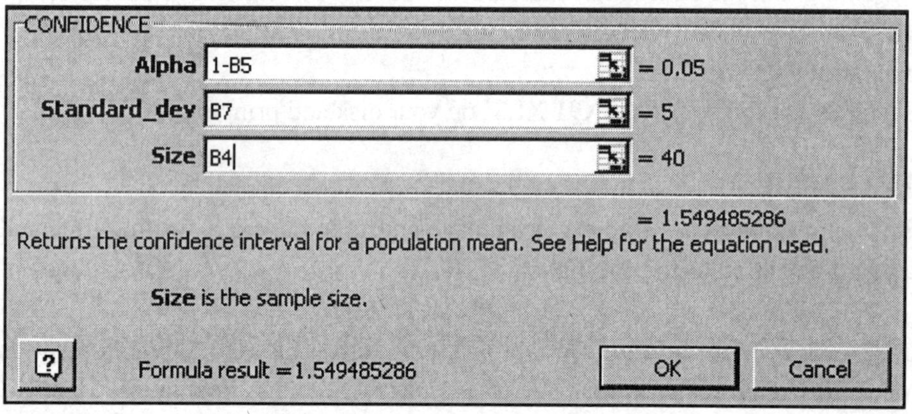

Method 2 In cell B9, type the formula **=CONFIDENCE(.05,5,40)** or **=CONFIDENCE(1-B5, B7, B4)** to obtain the value of B.

Method 3 Instead of using the **CONFIDENCE** function to obtain B, we can also use the *Function Name* **NORMSINV** under *Function Category* **Statistical** of the **Paste Function** in the following way. To input the probability for **NORMSINV**, first we need the cumulative probability for $z_{\alpha/2}$, which is 1 - (α/2) (see the diagram below). Since we have already stored 1 - α in cell B5, we have α = (1 - B5). Therefore, the required cumulative probability 1 - (α/2)

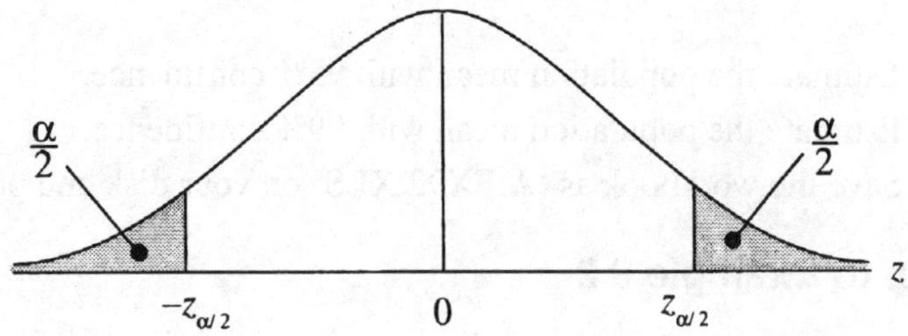

= 1 - ((1-B5)/2). In cell C5, type the label 'Cumulative probability, 1-(Alpha/2) =' and, in cell D5, type the formula **=1-((1-B5)/2)**. Now, to obtain B, type the explicit formula **=NORMSINV(D5)*B7/SQRT(B4)**, in cell B9.

Calculating the upper and lower limits of the confidence interval

Type the labels 'Lower limit =' in cell A11 and 'Upper limit =' in cell A12. To calculate the lower limit of the interval, in cell B11, type the formula **=B6-B9**. Type the formula **=B6+B9** in cell B12 to obtain the upper limit of the confidence interval. The resulting output will look like Figure 9.2.

Figure 9.2 Confidence interval for μ when σ is known

	A	B	C	D
1	95% Confidence Interval for the Single Population Mean Mu			
2	Population Standard Deviation Sigma is Known			
3				
4	Sample size n =	40		
5	Confidence level =	0.95	Cumulative probability, 1-(Alpha/2) =	0.975
6	Sample mean Xbar =	7.2		
7	Population SD sigma =	5		
8				
9	Half width B =	1.549485286	Half width B =	1.549485286
10				
11	Lower limit =	5.650514714		
12	Upper limit =	8.749485286		

Task 2 Save the workbook as **A:EX91.XLS** on your disk and print the file.

9.3 Example 9.2: Constructing confidence intervals for the population mean (μ) when σ is unknown

The following are the sample returns from an investment:

4, 8, 12, 11, 14, 6, 12, 8, 9 5.

If all returns for that investment come from a normal population,

Task 1 Estimate the population mean with 95% confidence.
Task 2 Estimate the population mean with 99% confidence.
Task 3 Save the workbook as 'A:EX92.XLS' on your disk and print the file.

Solution to Example 9.2

[In this example, we are given only the sample raw data set and we have to calculate the sample mean $\overline{X}$ and sample standard deviations from the data set. In some instances when we are not given the raw data set, we will be given the calculated sample mean and sample standard deviation values.]

Task 1 Open a new EXCEL workbook. Type the label 'Sample data' in cell A1 and type the sample data in cells B1 to B10. In cell A15, type heading '95% Confidence Interval for the Single Population Mean Mu' and centre across A15 to G15. Type 'Population Standard Deviation Sigma is Unknown' in cell A16 and centre across A16 to G16. Since the population standard deviation σ is *unknown*, to construct a 95% confidence interval for the population mean μ, we use the formula (Case 2, Section 9.1)

$$\overline{X} \pm \text{B, where B} = t_{\alpha/2,n-1}\frac{s}{\sqrt{n}} = t_{.025,9}\frac{s}{\sqrt{n}}$$

First, in cells A18, A19 , A20 and A21, type the labels 'Sample size n =', and 'Confidence level =', 'Sample mean Xbar =' and 'Sample SD s =' to input the given information. In cells B18 and B19, type the values **10** and **.95**. In cells B20 and B21, type the formulas **=AVERAGE($B$1:$B$10)** and **=STDEV($B$1:$B$10)** to obtain the mean and standard deviation of the sample data. [Note: Since we are going to use the same data set as in Task 2, to keep the data cell locations fixed, we use the $ format for the cell locations.] In cell A23, type the label 'SE of sample mean ='. In cell B23, type the formula **=B21/SQRT(B18)**.

Calculating half the width (B) of the confidence interval

In Example 9.1, when B involves with the normal distribution, we use the *Function Name* **CONFIDENCE** to obtain the value of B. No such facility is available when B does not involve with a normal distribution. In this example, since B involves the t-distribution, first we use EXCEL to obtain $t_{.025,9}$ and then write our own formula to calculate B.

Method 1 In cell A24, type the label 't-value =' and then click on cell B24. Now click on the icon **Paste Function** on the *Standard toolbar* and select the *Function Category* **Statistical** and *Function Name*

TINV and click **OK**. The TINV dialogue box will ask for two input values (see Figure 9.3): (1) *Probability* (which is the same as α); and (2) *Deg_freedom* (which is n-1). For a 95% confidence interval, input *Probability* as **.05** or **1-B19** (for a 90% CI the corresponding input is .10) and for a sample size of 10, input degrees of freedom as **9** or **B18-1** and then click **OK**. Now, the value for $t_{.025,9}$ will appear in cell B24. In cell A25, type the label 'Half width B ='. To calculate B, in cell B25, type the formula **=B23*B24**.

Figure 9.3 TINV Function dialogue box

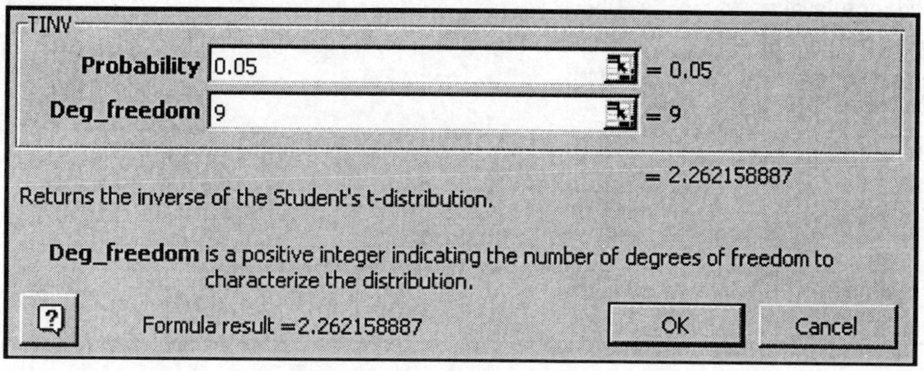

Method 2 In cell A24, type the label 't-value =' and then click on cell B24. Now type the function **=TINV(1-B19, B18-1)**, the value for $t_{.025,9}$ will appear in cell B24. In cell A25, type the label 'Half width B ='. To calculate B, in cell B25, type the formula **=B24*B23**.

Calculating the upper and lower limits of the confidence interval

Type the labels 'Lower limit =' in cell A27 and 'Upper limit =' in cell A28. To calculate the lower limit of the interval, in cell B27, we write the formula**=B20-B25**. Type the formula **=B20+B25** in cell B28 to obtain the upper limit of the confidence interval. The results are presented in Figure 9.4.

Task 2 To obtain the 99% confidence interval for the population mean, we follow the same procedure as in Task 1 above where we replace .90 with .99. The shortcut way to do this task is to copy the full contents of Task 1, highlight cell A15 to E28 and click on the icon **Copy** on the *Standard toolbar* and then click on cell A35. Now click on the icon **Paste** on the *Standard toolbar*. This will copy all the commands from Task 1 into cells A35 to E48. It can be seen that when copying, the relevant cell references in the formulas will all be automatically changed. Now change 90% in cell A35 to 99% and .90 in cell B39 to .99. EXCEL will automatically convert the relevant calculations. The new B value will be displayed in cell B45 and the upper and lower limits of the interval will be displayed in cells B47 and B48. The results are shown in Figure 9.5.

Figure 9.4 95% confidence interval for μ when σ is unknown

	A	B	C	D	E
15		95% Confidence Interval for a Single Population Mean Mu			
16		Population Standard Deviation Sigma is Unknown			
17					
18	Sample size n =	10			
19	Confidence level =	0.95			
20	Sample mean Xbar =	8.9			
21	Sample SD s =	3.314949			
22					
23	SE of Sample mean =	1.048279			
24	t-value =	2.262159			
25	Half width B =	2.371374			
26					
27	Lower limit =	6.528626			
28	Upper limit =	11.27137			

Figure 9.5 99% confidence interval for μ when σ is unknown

	A	B	C	D	E
35		99% Confidence Interval for the Single Population Mean Mu			
36		Population Standard Deviation Sigma is Unknown			
37					
38	Sample size n =	10			
39	Confidence level =	0.99			
40	Sample mean Xbar =	8.9			
41	Sample SD s =	3.314949			
42					
43	SE of Sample mean =	1.048279			
44	t-value =	3.249843			
45	Half width B =	3.406742			
46					
47	Lower limit =	5.493258			
48	Upper limit =	12.30674			

Task 3 Save the workbook as **A:EX92.XLS** on your disk and print the file.

9.4 Example 9.3: Constructing confidence intervals for the population mean (μ) when σ is unknown and n > 30

To determine the mean waiting time for his customers, a bank manager took a random sample of 40 customers and found that the mean waiting time was 7.2 minutes.

Task 1 Assuming that the population standard deviation is unknown and the sample standard deviation from the sample of 40 customers waiting time is 4 minutes, find the 95% confidence interval estimate of the mean waiting time for all of the bank customers. Assume that the population is normal.

Task 2 Save the workbook as 'A:EX93.XLS' on your disk and print the file.

Solution to Example 9.3

Let X be the waiting time for customers at the bank and, μ be the mean waiting time and σ be the standard deviation of the population.

Task 1 In cell A1, type heading '95% Confidence Interval for the Single Population Mean Mu' and centre across A1 to D1. Type 'Population Standard Deviation Sigma is Unknown and n>30' in cell A2 and centre across A2 to G2. When the population standard deviation σ is *unknown* and n=40>30, to construct a 95% confidence interval for the population mean μ, we use the formula (Case 3, Section 9.1)

$$\overline{X} \pm \text{B, where B} = z_{.025} \frac{s}{\sqrt{n}}$$

In cells A4, A5 , A6 and A7, type the labels 'Sample size n=', 'Confidence level =', 'Sample mean Xbar =' and 'Sample SD s =' to input the given information. In cells B4 to B7, type the respective values **40**, **.95**, **7.2** and **4**.

Calculating half the width (B) of the confidence interval

As in Example 9.1, we can now use any one of the three methods. We use Method 2 here as it is the simplest.

In cell A9, type the label 'Half width B='. In cell B9, type the formula **=CONFIDENCE(.05,4,40)** or **=CONFIDENCE(1-B5, B7, B4)** to obtain the value of B.

Calculating the upper and lower limits of the confidence interval

Type the labels 'Lower limit =' in cell A11 and 'Upper limit =' in cell A12. To calculate the lower limit of the interval, in cell B11, we type the formula=**B6-B9**. Type the formula **=B6+B9** in cell B12 to obtain the upper limit of the confidence interval. The resulting output will look like Figure 9.6.

Figure 9.6 Confidence interval for μ when σ is unknown and n > 30

	A	B	C	D
1	95% Confidence Interval for the Single Population Mean Mu			
2	Population Standard Deviation Sigma is Unknown and n > 30			
3				
4	Sample size n =	40		
5	Confidence level =	0.95	Cumulative probability, 1-(Alpha/2) =	0.975
6	Sample mean Xbar =	7.2		
7	Sample SD s =	4		
8				
9	Half width B =	1.239588		
10				
11	Lower limit =	5.960412		
12	Upper limit =	8.439588		
13				

Task 2 Save the workbook as **A:EX93.XLS** on your disk and print the file.

9.5 Example 9.4: Determining the required sample size for estimating μ

The manager of a firm was interested in determining the sample size necessary to estimate the population salary of employees in a firm within $20 when the population standard deviation of salary is $100.

Task 1 Determine the sample size if the manager wants to estimate with 99% confidence.

Task 2 Determine the sample size if the manager wants to estimate with 90% confidence.

Task 3 Save the workbook as 'A:EX94.XLS' on your disk and print the file.

Solution to Example 9.4

To find the required sample size for a given level of confidence when the data type is quantitative, we use the formula (see Section 9.1)

$$n = \frac{z_{\alpha/2}^2 \sigma^2}{B^2}$$

We are given B=20, $\sigma = 100$.

Task 1 Open a new EXCEL workbook. For this task 1- α = .99. In cell A1, type the heading 'Estimating the Required Sample Size: Data Type Quantitative' and center it across cells A1 to D1. In cells A3, A4 and A5 type labels 'Half width B =', 'Confidence Level =' and 'Population SD sigma =' and, in cells B3, B4 and

B5, type the inputs, **20**, **.99** and **100**. To obtain *n*, we use the *Function Name* **NORMSINV** under *Function Category* **Statistical** of the **Paste Function** in the following way. To input the probability for **NORMSINV**, first we need the cumulative probability for $Z_{\alpha/2}$, which is 1 - (α/2).Since we have already stored 1 - α in cell B4, we have α = (1 - B4). Therefore, the required cumulative probability 1 - (α/2) = 1 - ((1-B4)/2). In cell C4, type the label 'Cumulative probability, 1-(Alpha/2) =' and, in cell D4, type the formula **=1-((1-B4)/2)**. Now, to obtain *n*, in cell A7, type 'Required sample size n =' and, in cell B7, type the explicit formula **=(NORMSINV(D4)*NORMSINV(D4)*B5*B5) /(B3*B3)**. This will result in cell B7 the required sample size for the 99% confidence level. The results are shown in the first part of Figure 9.7.

Task 2 To obtain the required sample size *n* for the 90% confidence interval for the population mean, we follow the same procedure as in Task 1 above where we replace .99 with .90. The shortcut way to do this task is to copy the full contents of Task 1 and make the necessary changes. To do this first highlight cell A1 to D7 and click on the icon **Copy** on the *Standard toolbar* and then click on cell A10. Now click on the icon **Paste** on the *Standard toolbar*. This will copy all the commands from Task 1 into cells A10 to D16. It can be seen that when copying, the relevant cell references in the formulas will all be automatically changed. Now change .99 in cell B13 to .90. EXCEL will automatically convert the relevant calculations. The new required sample size *n* will be displayed in cell B16. The results are shown in the lower part of Figure 9.7.

Figure 9.7 Estimation of the required sample size for quantitative data

	A	B	C	D
1	Estimating the Required Sample Size: Data Type Quantitative			
2				
3	Half width B =	20		
4	Confidence level =	0.99	Cumulative probability, 1 - (Alpha/2) =	0.995
5	Population SD sigma =	100		
6				
7	Required sample size n =	165.87		
8				
9				
10	Estimating the Required Sample Size: Data Type Quantitative			
11				
12	Half width B =	20		
13	Confidence level =	0.9	Cumulative probability, (1- Alpha/2) =	0.95
14	Population SD sigma =	100		
15				
16	Required sample size n =	67.63853483		

Task 3 Save the workbook as **A:EX94.XLS** on your disk and print the file.

9.6 Constructing confidence intervals involving a single population: Qualitative data type

When the data type is qualitative, the problem objective is to estimate the population proportion p. We use the sample proportion $\hat{p}$ as the point estimator for the population proportion p. [$\hat{p}$ is an unbiased estimator for p.] The confidence interval for p is given by

$$\hat{p} \pm \mathrm{B}, \quad \text{where } \mathrm{B} = z_{\alpha/2}\sqrt{\frac{\hat{p}\hat{q}}{n}}$$

provided $n\hat{p} \geq 5$ and $n\hat{q} \geq 5$, where the probability $(1 - \alpha)$ is the level of confidence, n is the sample size and $\hat{q} = 1 - \hat{p}$.

For a given level of confidence, to obtain the required sample size, we use the formula

$$n = \frac{z^2_{\alpha/2}\hat{p}\hat{q}}{\mathrm{B}^2}$$

Since we haven't yet taken a sample, $\hat{p}$ and $\hat{q}$ are unknown. The best way to proceed is to select values of $\hat{p}$ and $\hat{q}$ that would give the largest value of the sample size n. Therefore, we use $\hat{p} = .5$ and $\hat{q} = .5$.

9.7 Example 9.5: Constructing confidence intervals for the population proportion *p*

In a random sample of 200 individuals tax returns audited, 30 were found to be incorrect.

Task 1 Estimate with 99% confidence, the true proportion of incorrect individual tax returns.

Task 2 Estimate with 90% confidence, the true proportion of incorrect individual tax returns.

Task 3 Save the workbook as 'A:EX95.XLS' on your disk and print the file.

Solution to Example 9.5

Let p be the proportion of incorrect individual tax returns.

Task 1 Open a new EXCEL workbook. In cell A1, type the heading '99% Confidence Intervals for the Single Population Proportion p' and centre across A1 to G1. To construct a 99% confidence interval for the single population proportion p, we use the formula (see Section 9.6)

$$\hat{p} \pm \mathrm{B}, \quad \text{where } B = z_{.005}\sqrt{\frac{\hat{p}\hat{q}}{n}}; \qquad \hat{q} = 1 - \hat{p}.$$

First, in cells A3, A4, A5 and A6, type the labels 'Sample size n =', 'Confidence level =', 'Sample Xhat =' and 'Sample proportion phat =' to input the given information. In cells B3 to B5 type the corresponding values **200**, **.99** and **30**. In cell B6, type the formula **=B5/B3** to obtain the $\hat{p}$ value. In cell A8, type the label 'SE of sample proportion phat =' and, in cell B8, type the formula **=SQRT(B6*(1-B6)/B3)**.

Calculating half the width (B) of the confidence interval

Now type the label 'Half width B =' in cell A9 and click on cell B9. To obtain B, we use the *Function Name* **NORMSINV** under *Function Category* **Statistical** of the **Paste Function** in the following way. To input the probability for **NORMSINV**, first we need the cumulative probability for $z_{\alpha/2}$, which is 1 - (α/2). Since we have already stored 1 - α in cell B4, we have α = (1 - B4). Therefore, the required cumulative probability 1 - (α/2) = 1 - ((1-B4)/2). In cell C4, type the label 'Cumulative probability, 1-(Alpha/2) =' and, in cell D4, type the formula **=1-((1-B4)/2)**. Now, to obtain B, in cell B9, type the explicit formula **=NORMSINV(D4)*B8**.

Calculating the upper and lower limits of the confidence interval

Type the labels 'Lower limit =' in cell A11 and 'Upper limit =' in cell A12. To calculate the lower limit of the interval, in cell B11, we write the formula**=B6-B9**. Type the formula **=B6+B9** in cell B12 to obtain the upper limit of the confidence interval. The results are shown in Figure 9.8.

Figure 9.8 99% confidence interval for *p*

	A	B	C	D
1	99% Confidence Interval for the Single Population Proportion p			
2				
3	Sample size n =	200		
4	Confidence level =	0.99	Cumulative probability, 1 - (Alpha/2) =	0.995
5	Sample Xhat =	30		
6	Sample proportion phat =	0.15		
7				
8	SE of sample proportion phat =	0.025249		
9	Half width B =	0.065037		
10				
11	Lower limit =	0.084963		
12	Upper limit =	0.215037		
13				

Task 2 To obtain the 90% confidence interval for the population proportion, we follow the same procedure as in Task 1 above where we replace .99 with .90. The shortcut way to do this task is to copy the full contents of Task 1, highlight cell

A1 to D12 and click on the icon Copy on the *Standard toolbar* and then click on cell A17. Now click on the icon Paste on the *Standard toolbar.* This will copy all the commands from Task 1 into cells A17 to D28. It can be seen that when copying, the relevant cell references in the formulas will all be automatically changed. Now change 99% in cell A17 to 90% and .99 in cell B20 to .90. EXCEL will automatically convert the relevant calculations. The new B value will be displayed in cell B25 and the upper and lower limits of the interval will be displayed in cells B27 and B28. The results are shown in Figure 9.9.

Figure 9.9 90% confidence interval for *p*

	A	B	C	D
16				
17	90% Confidence Interval for the Single Population Proportion p			
18				
19	Sample size n =	200		
20	Confidence level =	0.9	Cumulative probability, 1 - (Alpha/2) =	0.95
21	Sample Xhat =	30		
22	Sample proportion phat =	0.15		
23				
24	SE of sample proportion =	0.025249		
25	Half width B =	0.041531		
26				
27	Lower limit =	0.108469		
28	Upper limit =	0.191531		
29				

Task 3 Save the workbook as **A:EX95.XLS** on your disk and print the file.

9.8 Example 9.6: Determining the required sample size for estimating *p*

A marketing manager is in the process of deciding whether to introduce a new product. He has concluded that he needs to perform a market survey in which he asks a random sample of people whether they will buy the product.

Task 1 How large a sample should he take if he wants to estimate the proportion of people who will buy the product to within 3 percent, with 99% confidence?

Task 2 How large a sample should he take if he wants to estimate the proportion of people who will buy the product to within 3 percent, with 90% confidence?

Task 3 Save the workbook as 'A:EX96.XLS' on your disk and print the file.

Solution to Example 9.6

To determine the required sample size to estimate the population proportion p for a given confidence level when the data type is qualitative, we use the formula (see Section 9.6)

$$n = \frac{z_{\alpha/2}^2 \hat{p}\hat{q}}{B^2}$$

We are given B=3% = .03 and no information on $\hat{p}$ or $\hat{q}$. Therefore, to obtain the largest sample size, we use $\hat{p}$ =.5 and $\hat{q}$ =.5.

Task 1 For this task, 1- α = .99. In cells A1, type the heading 'Estimating the Required Sample Size: Data Type Qualitative, Alpha=.01' and center it across cells A1 to D1. In cells, A3, A4 and A5, type labels 'Half width B =', 'Confidence Level =' and 'Assumed sample proportion phat =' and, in cells B3, B4 and B5, type the input values, **.03**, **.99** and **.5**. To obtain *n*, we use the *Function Name* **NORMSINV** under *Function Category* **Statistical** of the **Paste Function** in the following way. To input the probability for **NORMSINV**, first we need the cumulative probability for $z_{\alpha/2}$, which is 1 - (α/2). Since we have already stored (1 - α) in cell B4, we have α = (1 - B4). Therefore, the required cumulative probability 1 - (α/2) = 1 - ((1-B4)/2). In cell C4, type the label 'Cumulative probability, 1-(Alpha/2) =' and, in cell D4, type the formula **=1-((1-B4)/2)**. Now, to obtain n, in cell A7, type 'Required sample size n =' and, in cell B7, type the explicit formula **=(NORMSINV(D4)*NORMSINV(D4)*B5*(1-B5))/(B3*B3)**. This will give the required sample size n for the 99% confidence level in cell B7. The results are shown in the first part of Figure 9.10.

Figure 9.10 Estimation of the required sample size for qualitative data

	A	B	C	D
1	Estimating the Required Sample Size: Data Type Qualitative, Alpha = .01			
2				
3	Half width B =	0.03		
4	Confidence level =	0.99	Cumulative probability, (1- Alpha/2)	0.995
5	Assumed sample proportion phat	0.5		
6				
7	Required sample size n =	1843.03		
8				
9				
10	Estimating the Required Sample Size: Data Type Qualitative, Alpha = .10			
11				
12	Half width B =	0.03		
13	Confidence level =	0.9	Cumulative probability (1 - Alpha/2)	0.95
14	Assumed sample proportion phat	0.5		
15				
16	Required sample size n =	751.5393		

Task 2 To obtain the required sample size *n* for the 90% confidence interval for the population proportion *p*, we follow the same procedure as in Task 1 above where we replace .99 with .90. The shortcut way to do this task is to copy the

full contents of Task 1 and make the necessary changes. To do this first highlight cell A1 to D7 and click on the icon **Copy** on the *Standard toolbar* and then click on cell A10. Now click on the icon **Paste** on the *Standard toolbar*. This will copy all the commands from Task 1 into cells A10 to D16. It can be seen that when copying, the relevant cell references in the formulas will all be automatically changed. Now change the .01 in the heading in cell A10 to .10 and change .99 in cell B13 to .90. EXCEL will automatically convert the relevant calculations. The required sample size *n* for the 90% confidence level will be displayed in cell B16. The results are shown in the lower part of Figure 9.10.

Task 3 Save the workbook as **A:EX96.XLS** on your disk and print the file.

9.9 Constructing confidence intervals involving two populations: Quantitative data type

When making inferences about two populations where the data type is quantitative, the problem objective is to compare the population means. The parameter of interest is the difference between the population means ($\mu_1 - \mu_2$). We use the difference between the two sample means ($\overline{X}_1 - \overline{X}_2$) as the point estimator for ($\mu_1 - \mu_2$).

Case 1 ***When both population variances σ_1^2 and σ_2^2 are known***

The (1 - α)100% confidence interval for $\mu_1 - \mu_2$ is given by

$$(\overline{X}_1 - \overline{X}_2) \pm B, \quad \text{where } B = z_{\alpha/2}\sqrt{\frac{\sigma_1^2}{n_1} + \frac{\sigma_2^2}{n_2}}$$

provided $\overline{X}_1$ and $\overline{X}_2$ are normally distributed, where the probability (1 - α) is the level of confidence and n_1 *and* n_2 are sample sizes. Note that the sample means are normally distributed if either both populations are normally distributed or if the distribution of the populations are unknown or non-normal and the sample sizes $n_1, n_2 \geq 30$ (using the Central Limit theorem).

Case 2 ***When the population variances σ_1^2 and σ_2^2 are unknown but are equal***

The confidence interval for ($\mu_1 - \mu_2$) is given by

$$(\overline{X}_1 - \overline{X}_2) \pm B, \quad \text{where } B = t_{\alpha/2,\nu}\sqrt{\frac{s_P^2}{n_1} + \frac{s_P^2}{n_2}}$$

provided both populations are normally distributed, where the probability (1 - α) is the level of confidence, $\nu = n_1 + n_2 - 2$ is the degrees of freedom of the t-distribution and **Pooled variance**

$$s_P^2 = \frac{(n_1-1)s_1^2 + (n_2-1)s_2^2}{n_1+n_2-2}$$

Case 3 ***When the population variances σ_1^2 and σ_2^2 are unknown and sample sizes $n_1, n_2 > 30$***

The (1 - α)100% confidence interval for $(\mu_1 - \mu_2)$ is given by

$$(\overline{X}_1 - \overline{X}_2) \pm \text{B}, \quad \text{where } B = t_{\alpha/2,d.f}\sqrt{\frac{s_1^2}{n_1}+\frac{s_2^2}{n_2}} \approx z_{\alpha/2}\sqrt{\frac{s_1^2}{n_1}+\frac{s_2^2}{n_2}}$$

provided both populations are normally distributed and *the sample sizes n_1, n_2 > 30*, where the probability (1 - α) is the level of confidence. Here we use the z-value, since for large sample sizes, the t-distribution can be approximated by the standard normal distribution.

9.10 Example 9.7: Constructing confidence intervals for the difference between two population means ($\mu_1 - \mu_2$) when σ_1 and σ_2 are known

In order to compare the wages paid to workers in two large companies, random samples of wage earners are drawn from each population. The average hourly wages are $\overline{X}_1$ = \$7.50 and $\overline{X}_2$=\$7.90. Estimate the difference in average wages between the two companies with 95% confidence.

Task 1 If samples of size 20 each were taken and the population standard deviations are known and are σ_1 = \$2.00 and σ_2 = \$1.80. (Assume both populations are normally distributed.)

Task 2 Save your workbook as 'A:EX97.XLS' on your disk and print the file.

Solution to Example 9.7

Task 1 In cell A1, type heading '95% Confidence Interval for the Difference between Two Population Means Mu1-Mu2'. Type heading 'Population Standard Deviations Sigma1 and Sigma2 are known' in cell A2. When the population standard deviations are *known*, to construct a 95% confidence interval for the difference of population means, we use the formula (Case 1, Section 9.9)

$$\left(\overline{X}_1 - \overline{X}_2\right) \pm \text{B, where } B = z_{.025}\sqrt{\frac{\sigma_1^2}{n_1}+\frac{\sigma_2^2}{n_2}}\,.$$

First, in cell A4, type the label 'Confidence level =' and, in cell B4, type the value **.95**. In cells A6 and C6, type labels 'Sample 1' and 'Sample 2'. In cells A8, A9 and A10, type the labels 'Sample size n1 =', 'Sample mean Xbar1 =' and 'Population SD sigma1 =' to input the given information for Sample 1. In cells B8 to B10, type the corresponding values **20**, **7.50** and **2.00**. Similarly, in cells C8, C9 and C10, type the labels 'Sample size n2 =', 'Sample mean Xbar2 =' and 'Population SD sigma2 =' for Sample 2. In cells D8 to D10, type the corresponding values **20**, **7.90** and **1.80**.

Calculating half the width (B) of the confidence interval

In cell A12, type the label 'Standard error of Xbar1-Xbar2 =' and, in cell B12, type the formula **=SQRT((B10*B10/B8)+(D10*D10/D8))**. Now type 'Half width B =' in cell A13. To obtain B, we use the *Function Name* **NORMSINV** under *Function Category* **Statistical** of the **Paste Function** in the following way. To input the probability for **NORMSINV**, first we need the cumulative probability for $z_{\alpha/2}$, which is 1 - (α/2). Since we have already stored (1 - α) in cell B4, we have α = (1 - B4). Therefore, the required cumulative probability 1 - (α/2) = 1 - ((1-B4)/2). In cell C4, type the label 'Cumulative probability, 1-(Alpha/2) =' and, in cell D4, type the formula **=1-((1-B4)/2)**. Now, to calculate B, in cell B13, type the formula **=NORMSINV(D4)* B12**.

Calculating the upper and lower limits of the confidence interval

Type the labels 'Lower limit =' in cell A15 and 'Upper limit =' in cell A16. To calculate the lower limit of the interval, in cell B15, we write the formula**=B9-D9-B13**. Type the formula **=B9-D9+B13** in cell B16 to obtain the upper limit of the confidence interval. The resulting output will look like the following in Figure 9.11.

Figure 9.11 Confidence interval for μ_1 - μ_2 when σ_1 and σ_2 are known

	A	B	C	D
1	95% Confidence Interval for the Difference between Two Population Means Mu1-Mu2			
2	Population Standard Deviations Sigma1 and Sigma2 are Known			
3				
4	Confidence level =	0.95	Cumulative probability, 1 -(Alpha/2)	0.975
5				
6	Sample 1		Sample 2	
7				
8	Sample size n1 =	20	Sample size n2 =	20
9	Sample mean Xbar1 =	7.5	Sample mean Xbar2 =	7.9
10	Population SD sigma1 =	2	Population SD sigma2 =	1.8
11				
12	Standard error of Xbar1-Xbar2	0.60166		
13	Half width B =	1.17924		
14				
15	Lower limit =	-1.5792		
16	Upper limit =	0.77924		

Task 2 Save the workbook as **A:EX97.XLS** on your disk and print the file.

9.11 Example 9.8: Constructing confidence intervals for the difference between two population means (μ_1 - μ_2) when σ_1 and σ_2 are unknown but equal

To compare the mean income of two suburbs in Brisbane the following random samples were drawn from each suburb (income in $'000).

<u>Sample 1</u> 14, 29, 32, 18, 24
<u>Sample 2</u> 41, 36, 40, 27, 23, 32, 37.

Task 1 Calculate the mean and standard deviation of the two samples.

Task 2 If the population variances of the two suburbs are unknown but are considered to be equal, estimate the difference between their population mean income, with 95% confidence.

Task 3 Save the workbook as 'A:EX98.XLS' on your disk and print the file.

Solution to Example 9.8

Task 1 Open a new EXCEL workbook. Type the labels 'Sample 1' in cell A1 and 'Sample 2' in cell C1 and the data for sample 1 in cells B3 to B7 and for sample 2 in cell D3 to D9. In cells A11, A12 and A13, type the labels 'Sample size n1 =', 'Sample mean Xbar1 =' and 'Sample SD s1 =' and in cells C11, C12 and C13, type the labels 'Sample size n2 =', 'Sample mean Xbar2 =' and 'Sample SD s2 =' to calculate the required information for Samples 1 and 2. In cells B11 and D11, type the sample sizes **5** and **7** and in B12, B13 and D12 and D13, type the formulas **=AVERAGE(B3:B7)**, **=STDEV(B3:B7)** and formulas **=AVERAGE(D3:D9)**, **=STDEV(D3:D9)** to obtain the mean and standard deviation of Sample 1 and Sample 2, respectively.

Task 2 In cell A15, type heading '95% Confidence Intervals for the Difference between Two Population Means Mu1-Mu2'. Type heading 'Population Standard Deviations are Unknown but Equal' in cell A16. When the population standard deviations are *<u>unknown but equal</u>*, to construct a 95% confidence interval for the difference of population means, we use the formula (<u>Case 2, Section 9.9</u>)

$$\left(\overline{X}_1 - \overline{X}_2\right) \pm \text{B, where } B = t_{.025,d.f.}\sqrt{\frac{s_P^2}{n_1} + \frac{s_P^2}{n_2}},$$

where d.f. = n_1+ n_2-2 and **Pooled variance**

$$s_P^2 = \frac{(n_1 - 1)s_1^2 + (n_2 - 1)s_2^2}{n_1 + n_2 - 2}$$

First, in cell A18, type the label 'Confidence level =' and, in cell B18, type the corresponding value **.95**.

Calculating half the width (B) of the confidence interval

In cell A20, type the label 'Degrees of Freedom =' and in cell B20, type the formula **=B11+D11-2**. In cell A21, type the label 'Pooled Variance =' and, in cell B21, type the formula **=(((B11-1)*B13*B13)+((D11-1)*D13*D13))/(B11+D11-2)**. In cell A22, type the label 'Standard error of Xbar1-Xbar2 =' and, in cell B22, type the formula **=SQRT((B21/B11)+(B21/D11))**. Now type 'Half width B=' in cell A23. To calculate B, in cell B23, type the formula **=TINV(1-B18,B20)*B22**.

Calculating the upper and lower limits of the confidence interval

Type the labels 'Lower limit =' in cell A25 and 'Upper limit =' in cell A26. To calculate the lower limit of the interval, in cell B25, type the formula **=B12-D12-B23**. Type the formula **=B12 -D12+B23** in cell B26 to obtain the upper limit of the confidence interval. The resulting output will look like the following in Figure 9.12.

Figure 9.12 Confidence interval for $\mu_1 - \mu_2$ when σ_1 and σ_2 are unknown and $\sigma_1 = \sigma_2$

	A	B	C	D
11	Sample size n1 =	5	Sample size n2 =	7
12	Sample mean Xbar1 =	23.4	Sample mean Xbar2 =	33.71429
13	Sample SD s1=	7.469939759	Sample SD s2=	6.725927
14				
15	95% Confidence Interval for the Difference between Two Population Means Mu1-Mu2			
16	Population Standard Deviations are Unknown but Equal			
17				
18	Confidence level =	0.95		
19				
20	Degrees of freedom =	10		
21	Pooled variance =	49.46285714		
22	Standard error of Xbar1-Xbar2 =	4.118093476		
23	Half width B =	9.175685658		
24				
25	Lower limit =	-19.4899714		
26	Upper limit =	-1.13860006		

Task 3 Save the workbook as **A:EX98.XLS** on your disk and print the file.

9.12 Example 9.9: Constructing confidence intervals for the difference between two population means ($\mu_1 - \mu_2$) when σ_1 and σ_2 are unknown and n_1, n_2 > 30

In order to compare the wages paid to workers in two large companies, random samples of wage earners are drawn from each population. The average hourly wages are $\overline{X}_1 =$

\$7.50 and $\overline{X}_2$=\$7.90. Estimate the difference in average wages between the two companies with 95% confidence.

Task 1 If samples of size 50 each were taken and the population standard deviations are unknown and the sample standard deviations are s_1 = \$1.70 and s_2=\$2.20.

Task 2 Save the workbook as 'A:EX99.XLS' on your disk and print the file.

Solution to Example 9.9

Task 1 In cell A1, type heading '95% Confidence Interval for the Difference between Two Population Means Mu1-Mu2'. Type heading 'Population Standard Deviations are Unknown and Sample Sizes are > 30' in cell A2. When the population standard deviations are *unknown and the sample sizes n_1, n_2 > 30*, to construct a 95% confidence interval for $\mu_1 - \mu_2$ we use the formula (see Case 3, Section 9.9)

$$\left(\overline{X}_1 - \overline{X}_2\right) \pm \text{B, where } B = z_{.025}\sqrt{\frac{s_1^2}{n_1} + \frac{s_2^2}{n_2}}.$$

First, in cell A4, type the label 'Confidence level =' and in cell B4 type the value **.95**. In cells A6 and C6 type labels 'Sample 1' and 'Sample 2'. In cells A8, A9 and A10 type the labels 'Sample size n1 =', 'Sample mean Xbar1 =' and 'Sample SD s1 =' to input the given information for Sample 1. In cells B8 to B10 type the corresponding values **50**, **7.50** and **1.70**. In cells C8, C9 and C10, type the labels 'Sample size n2 =', 'Sample mean Xbar2 =' and 'Sample SD s2 =' to input the given information for Sample 2. In cells D8 to D10, type the corresponding values **50, 7.90** and **2.20**.

Calculating half the width (B) of the confidence interval

In cell A12, type the label 'Standard Error of Xbar1-Xbar2 =' and, in cell B12, type the formula **=SQRT((B10*B10/B8)+(D10*D10/D8))**. Now type 'Half width B =' in cell A13. To obtain B, as above we use the *Function Name* **NORMSINV** under *Function Category* **Statistical** of the **Paste Function.** In cell C4, type the label 'Cumulative probability, 1-(Alpha/2) =' and, in cell D4, type the formula **=1-((1-B4)/2)**. Now, to calculate B, in cell B13, type the formula **=NORMSINV(D4)* B12.**

Calculating the upper and lower limits of the confidence interval

Type the labels 'Lower limit =' in cell A15 and 'Upper limit =' in cell A16. To calculate the lower limit of the interval, in cell B15, we write the formula**=B9-D9-B13**. Type the formula **=B9-D9+B13** in cell B16 to obtain the upper limit of the confidence interval. The resulting output will look like the following in Figure 9.13.

Figure 9.13 Confidence interval for μ_1 - μ_2 when σ_1 and σ_2 unknown and n_1, n_2 >30

	A	B	C	D
1	95% Confidence Interval for the Difference between Two Population Means Mu1-Mu2			
2	Population Standard Deviations are Unknown and Sample Sizes are >30			
3				
4	Confidence level =	0.95	Cumulative probability, 1 -(Alpha/2)	0.975
5				
6	Sample 1		Sample 2	
7				
8	Sample size n1 =	50	Sample size n2 =	50
9	Sample mean Xbar1 =	7.5	Sample mean Xbar2 =	7.9
10	Sample SD s1=	1.7	Sample SD s2=	2.2
11				
12	Standard error of Xbar1-Xbar2	0.393192		
13	Half width B =	0.770641		
14				
15	Lower limit =	-1.17064		
16	Upper limit =	0.370641		

Task 2 Save your file with filename **A:EX99.XLS** and print the file.

9.13 Constructing confidence intervals involving two populations: Qualitative data type

When making inferences about two populations where the data type is qualitative, the problem objective is to compare the population proportions. The parameter of interest is the difference between the population proportions (p_1 - p_2). We use the difference between the two sample proportions $(\hat{p}_1 - \hat{p}_2)$ as the point estimator for (p_1 - p_2). The confidence interval for (p_1 - p_2) is given by

$$(\hat{p}_1 - \hat{p}_2) \pm \mathrm{B}, \qquad \text{where } \mathrm{B} = z_{\alpha/2}\sqrt{\frac{\hat{p}_1\hat{q}_1}{n_1} + \frac{\hat{p}_2\hat{q}_2}{n_2}}$$

provided $n_1\hat{p}_1 \geq 5$, $n_1\hat{q}_1 \geq 5$, $n_2\hat{p}_2 \geq 5$ and $n_2\hat{q}_2 \geq 5$, where the probability (1 - α) is the level of confidence, n_1 *and* n_2 are sample sizes, $\hat{q}_1 = 1 - \hat{p}_1$ and $\hat{q}_2 = 1 - \hat{p}_2$.

9.14 Example 9.10: Constructing confidence intervals for the difference between two population proportions (p_1-p_2)

A market researcher, employed by a chain of service centers offering no-wait oil and filter changes, wants to know the difference in the fraction of male and female car owners who regularly use the service. Such information will be quite useful in designing advertising. In a random sample of 500 men, 42 indicated that they frequently have their cars serviced by this chain. A random sample of 300 women showed that 38 use the service.

Task 1 Estimate the difference in the proportion of men and women who use the oil-change service with 90% confidence.

Task 2 Estimate the difference in the proportion of men and women who use the oil-change service with 99% confidence.

Task 3 Save the workbook as 'A:EX910.XLS' on your disk and print the file.

Solution to Example 9.10

Task 1 In cell A1, type heading '90% Confidence Interval for the Difference between Two Population Proportions p1-p2' and centre across A1 to G1. To construct a 90% confidence interval for the difference of population proportions, we use the formula (see Section 9.13)

$$(\hat{p}_1 - \hat{p}_2) \pm \mathrm{B}, \qquad \text{where } B = z_{.025}\sqrt{\frac{\hat{p}_1\hat{q}_1}{n_1} + \frac{\hat{p}_2\hat{q}_2}{n_2}}.$$

First, in cells A3, type the label 'Confidence level =' and in cell B3 type its value '.90'. In cells A5, A6 and A7, type the labels 'Sample size n1 =', 'Sample Xhat1 =' and 'Sample proportion phat1 =' and, in cells C5, C6 and C7, type the labels 'Sample size n2 =', 'Sample Xhat2 =' and 'Sample proportion phat2 =' to input the given information for sample 1 and sample 2, respectively. In cells B5 to B6 and D5 to D6, type the corresponding values **500** and **42**, and **300** and **38**. In cells B7 and D7, type the formula **=B6/B5** and **=D6/D5** to obtain the $\hat{p}_1$ and $\hat{p}_2$ values.

Calculating half the width (B) of the confidence interval

In cell A9, type the label 'Standard Error of phat1-phat2 =' and, in cell B9, type the formula **=SQRT((B7*(1-B7)/B5)+(D7*(1-D7)/D5))**. Now type 'Half Width B=' in cell A10. In cell C3, type the label 'Cumulative probability, 1-(Alpha/2) =' and, in cell D3, type the formula **=1-((1-B3)/2)**. To calculate B, in cell B10, type the formula **=NORMSINV(D3)* B9**.

Calculating the upper and lower limits of the confidence interval

Type the labels 'Lower limit =' in cell A12 and 'Upper limit =' in cell A13. To calculate the lower limit of the interval, in cell B12, type the formula**=B7-D7-B10**. Type the formula **=B7-D7+B10** in cell B13 to obtain the upper limit of the confidence interval. The resulting output is shown in the first part of Figure 9.15.

Task 2 To obtain the 99% confidence interval for the difference between the population proportions, we follow the same procedure as in Task 1 above where we replace .90 with .99. The shortcut way to do this task is to copy the full contents of Task

1, highlight row 1 to row 13 and click on the icon **Copy** on the *Standard toolbar* and then click on cell A18. Now click on the icon **Paste** on the *Standard toolbar*. This will copy all the commands from Task 1 into rows 18 to 30. It can be seen that when copying, the relevant cell references in the formulas will all be automatically changed. Now change 90% in cell A18 to 99% and .90 in cell B20 to .99. EXCEL will automatically convert the relevant calculations. The new B value will be displayed in cell B27 and the upper and lower limits of the interval will be displayed in cells B29 and B30. The resulting output is shown in the lower part of Figure 9.14.

Figure 9.14 Confidence intervals for p_1 - p_2

	A	B	C	D
1	90% Confidence Interval for the Difference between Two Population Proportions p1-p2			
2				
3	Confidence level =	0.9	Cumulative probability, 1-(Alpha/2)	0.95
4				
5	Sample size n1 =	500	Sample size n2 =	300
6	Sample Xhat1 =	42	Sample Xhat2 =	38
7	Sample proportion phat1 =	0.084	Sample proportion phat2 =	0.1266667
8				
9	Standard error of phat1-phat2	0.02286		
10	Half width B =	0.0376		
11				
12	Lower limit =	-0.0803		
13	Upper limit =	-0.0051		
14				
15				
16				
17				
18	99% Confidence Interval for the Difference between Two Population Proportions p1-p2			
19				
20	Confidence level =	0.99	Cumulative probability, 1-(Alpha/2)	0.995
21				
22	Sample size n1 =	500	Sample size n2 =	300
23	Sample Xhat1 =	42	Sample Xhat2 =	38
24	Sample proportion phat1 =	0.084	Sample proportion phat2 =	0.1266667
25				
26	Standard error of phat1-phat2	0.02286		
27	Half width B =	0.05889		
28				
29	Lower limit =	-0.1016		
30	Upper limit =	0.01622		

Task 3 Save the workbook as **A:EX910.XLS** on your disk and print the file.

9.15 Exercises to Chapter 9

The following exercises are based on exercises from Chapters 8 and 9 of Selvanathan et al., Australian Business Statistics, 1994.

Exercise 9.1

A survey of 20 Australian companies indicated that the average annual income of company chief executives was $120,000.

Task 1 Assuming that the population standard deviation is $7,500 and that the annual incomes are normally distributed, calculate the 90% confidence interval estimate of the average annual income of all company chief executives.

Task 2 Save your workbook as 'A:EXEC.XLS' and print the file.

Exercise 9.2

The following data were drawn from a normal population:

4, 8, 12, 11, 14, 6, 12, 8, 9, 5.

Task 1 Estimate the population mean with 90% confidence.

Task 2 Save your workbook as 'A:NORM.XLS' and print the file.

Exercise 9.3

An advertisement for a major washing machine manufacturer in Sydney claims that its repairmen are the loneliest in the world because its washing machines require the smallest number of service calls. To examine this claim, researchers drew a random sample of 100 owners of 5-year-old washing machines of that manufacturer. The mean and the standard deviation of the number of service calls in the 5-year period were 4.3 and 1.8, respectively.

Task 1 Find the 90% confidence interval estimate for the mean number of service calls for all 5-year-old washing machines produced by that manufacturer.

Task 2 Save your workbook as 'A:WASH.XLS' and print the file.

Exercise 9.4

A manufacturer of a brand of designer jeans in Melbourne realizes that many retailers charge less than the suggested retail price of $40. A random sample of 20 retailers reveals that the mean and the standard deviation of the prices of the jeans are $32 and $2.50, respectively.

Task 1 Estimate with 90% confidence the mean retail price of the jeans.

Task 2 Save your workbook as 'A:JEANS.XLS' and print the file.

Exercise 9.5

A department store wants to estimate the average account of all its credit-card customers, to within $10.00, with 99% confidence. A quick analysis tells us that the smallest account shows $0 while the largest is $500.

Task 1 Estimate the standard deviation.

Task 2 Determine the sample size.

Task 3 Save your workbook as 'A:CARD.XLS' and print the file.

Exercise 9.6

In a survey of 250 voters prior to election, 40% indicated that they would vote for the incumbent candidate.

Task 1 Estimate with 90% confidence the population proportion of voters who support the incumbent.

Task 2 Save your workbook as 'A:VOTE.XLS' and print the file.

Exercise 9.7

A firm's management is contemplating modifying one of its products. To help in making a decision, management representatives wish to conduct a market survey that will enable them to estimate the proportion of potential customers who would buy the new product. They wish to estimate this proportion to within three per cent, with 99% confidence.

Task 1 How large a sample should be drawn?

Task 2 Save your workbook as 'A:SURVEY.XLS' and print the file.

Exercise 9.8

In order to help make a financial decision, an investor observes 25 returns on one type of investment and 35 returns on a second type of investment. The sample means are $\overline{X}_1$ = 12.5 and $\overline{X}_2$ = 11.3. Assume that the returns are normally distributed, with standard deviations $\sigma_1 = \sigma_2 = 5$.

Task 1 Estimate the difference in mean returns, with 95% confidence.

Task 2 Save your workbook as 'A:RETURN.XLS' and print the file.

Exercise 9.9

Three years ago, a 100-acre site was planted with 250,000 white spruce seedlings. Half the site was scarified (soil turned up) and the seedlings spot-fertilized; the other half was not fertilized. Random samples of 50 trees from each half of the site were taken and the foliage of each tree was weighed. The results are shown in the accompanying table.

Fertilized Trees	Unfertilized Trees
$\overline{X}_1 = 5.98$	$\overline{X}_2 = 4.79$
$s_1^2 = .61$	$s_2^2 = .77$

Task 1 Estimate with 95% confidence the difference in mean foliage weight between fertilized and unfertilized trees.

Task 2 Save your workbook as 'A:FERT.XLS' and print the file.

Exercise 9.10

A major automobile manufacturer has plants in a number of countries. To examine productivity in Australia and Korea, a statistician counted the daily output for one randomly selected plant in Australia and a similar-sized plant in Korea. The mean and the standard deviation of the daily number of cars produced in 15 days were computed. These statistics are shown in the accompanying table.

Daily Car Production

Australia	**Korea**
$\overline{X}_1 = 325$	$\overline{X}_2 = 345$
$s_1 = 12$	$s_2 = 15$

Task 1 Estimate the difference in the mean number of cars produced between the plants in Australia and Korea. Use a confidence level of 99%.

Task 2 Save your workbook as 'A:KOREA.XLS' and print the file.

Exercise 9.11

Until recently, most car salespeople have been men. However, in the past decade, numerous automobile dealers have hired women as salespeople in the hope that they will sell more cars to female customers. A researcher decided to examine how successful female salespeople are. She took a random sample of eight female and eight male salespeople and recorded their commission income for the preceding year. The results are shown in the accompanying table.

Commission Income (in $1,000s)

	Male Salespeople	Female Salespeople
	31	35
	12	27
	52	24
	51	22
	20	55
	19	49
	28	14
	29	44
Mean	30.25	33.75
Variance	210.79	208.50

Task 1 Estimate with 95% confidence the difference in mean commission income between all male and female automobile salespeople.

Task 2 Save your workbook as 'A:MEN.XLS' and print the file.

Exercise 9.12

Physicians have been encouraging their patients to reduce smoking for many years. In a study to determine who smokes regularly, a random sample of 1,000 people were interviewed. Each was asked whether or not they smoke regularly, as well as a variety of questions relating to demographic characteristics (such as age and education). It was found that 248 respondents were university graduates. Of these, 52 smoked regularly. Of the remaining 752 respondents, 226 smoked regularly.

Task 1 Estimate with 90% confidence the difference in the fraction of smokers between university graduates and nongraduates.

Task 2 Save your workbook as 'A:DOCT.XLS' and print the file.

Chapter 10

Hypothesis Testing: Describing a Single Population

The previous chapter looked at making inferences using estimation of unknown population parameters. In this chapter, we use statistical inference to test hypotheses about population parameters. A *hypothesis* is a statement about the value that the population parameter can take. The purpose of hypothesis testing is to determine whether there exists enough evidence to conclude that a claim about a population parameter is reasonable. In this chapter, we test hypotheses about the same population parameters we considered in the previous chapter. The tests in this chapter are *parametric tests* as they test the value of the population parameter.

10.1 Definition of terms and the step-wise process used in hypothesis testing

There are two types of hypotheses, the null hypothesis (H_o) and the alternative hypothesis (H_A). The *null hypothesis* specifies that the population parameter is equal to a single value. The *alternative hypothesis* answers the question by specifying that the parameter is either 'greater than', 'less than' or 'not equal to' the specified value in the null hypothesis.

A *test statistic* of a test is a sample statistic upon which we base our decision to *reject* or *not to reject* a null hypothesis. The *decision rule* specifies the set of values of the test statistic for which the null hypothesis H_o is rejected in favour of H_A.

A *rejection region (or critical region)* consists of all the values of the test statistic for which the null hypothesis H_o is rejected in favour of the alternative H_A. The *acceptance region* consists of all the values of the test statistic for which the null hypothesis is not rejected. The value(s) that separates the rejection region(s) from the acceptance region is (are) called *critical value(s)*.

If the alternative is specified as 'not equal to', then the distribution of the test will have two rejection regions (one on each tail) and is called a *two-tail or two-sided test.* If the alternative is 'less than', then the rejection region will be on the left-hand tail and the test is called a *one-tail (left)* or *one-sided (left) test.* If the alternative is 'greater than', then the rejection region will be on the right-hand tail and the test is called a *one-tail (right)* or *one-sided (right) test.*

In a test, there are four possible outcomes:

1. Reject H_o when H_o is true - Incorrect decision
2. Reject H_o when H_o is false - Correct decision
3. Do not reject H_o when H_o is true - Correct decision
4. Do not reject H_o when H_o is false - Incorrect decision

The incorrect decision of *Rejecting H_o when H_o is true* is called making a *Type I error* and the incorrect decision of *not rejecting H_o when H_o is false* is called making a *Type II error*. The probability of a Type I error is denoted by α which is also called the *level of significance* and the probability of a Type II error is denoted by β. The decision rule is based on specifying the value of the level of significance α.

The *p-value (prob-value)* of a test is defined as the smallest value of α that would lead to the rejection of the null hypothesis. The power of a test is the probability that the test correctly rejects the null hypothesis whenever it is false. In other words, using the properties of complementary events, the power of a test is 1 minus the probability of not rejecting H_o when H_o is false, which is $(1-\beta)$.

There are two ways of testing a hypothesis. In both ways, we use a systematic approach by following a ***six-step process***. Note that all steps except Steps 4 and 6 are the same in both methods.

Method A ***Using the value of the test statistic***

Step 1 Set up the null and alternative hypotheses [since the alternative hypothesis answers the question, set this one up first] and based on the alternative, determine whether the test is (left) one-tail or (right) one-tail or two-tail.

Step 2 Determine the preferred test statistic and the sampling distribution of the standardised value of the test statistic.

Step 3 Specify the significance level. [We usually set α = .01, .05, .10, but other values are also possible.]

Step 4 Define the decision rule. [This involves the appropriate use of the statistical tables or the use of the appropriate function in EXCEL to determine the critical value(s) and the rejection region(s).]

Step 5 Calculate the observed value of the test statistic under the null hypothesis.

Step 6 Comparing the observed value of the test statistic obtained in Step 5 with the decision rule in Step 4, make a decision and answer the question.

Method B ***Using the p-value***

Step 1 Set up the null and alternative hypotheses [since the alternative hypothesis answers the question, set this one up first] and based on the alternative, determine whether the test is (left) one-tail or (right) one-tail or two-tail.

Step 2 Determine the preferred test statistic and the sampling distribution of the standardised value of the test statistic.

Step 3 Specify the significance level. [We usually set α = .01, .05, .10, but other values are also possible.]

Step 4 Define the decision rule in terms of α. That is, if p-value < α, reject H_o.

Step 5 Calculate the observed value of the test statistic under the null hypothesis and use this value to calculate the p-value.

Step 6 Compare the p-value obtained in Step 5 with the decision rule in Step 4 and make a decision and answer the question.

Aim: At the completion of the examples in this Chapter, you should be able to use EXCEL to perform hypothesis testing involving a single population about the

(i) population mean μ when the population variance σ^2 is known;

(ii) population mean μ when the population variance σ^2 is unknown; and

(iii) population proportion p.

10.2 Hypothesis testing about the population mean μ: Single population

When the data type is quantitative, the problem objective is to test hypotheses about the population mean μ. We follow the six-step procedure described above.

Step 1 The null and the alternative hypotheses will take the form:

H_o: $\mu = \mu_o$

H_A: $\mu \neq \mu_o$ (two-tail)

or H_A: $\mu < \mu_o$ (left) one-tail

or H_A: $\mu > \mu_o$ (right) one-tail

where μ_o is the value of the population mean μ under Ho.

Step 2 The test statistic takes the form:

Case 1 ***When the population variance σ^2 is known***

$$Z = \frac{\overline{X} - \mu}{\sigma/\sqrt{n}} \sim z(0,1)$$

provided either the population is normally distributed or if the distribution is unknown or non-normal and the sample size $n \geq 30$ (using the Central Limit theorem).

Case 2 ***When the population variance σ^2 is unknown***

$$t = \frac{\overline{X} - \mu}{s/\sqrt{n}} \sim t_{n-1}$$

provided the population is normally distributed.

Case 3 ***When the population variance σ^2 is unknown and the sample size $n > 30$***

$$t = \frac{\overline{X} - \mu}{s/\sqrt{n}} \sim t_{n-1} \approx z(0,1)$$

provided the population is normally distributed

Note that the difference between Cases 2 and 3 is that, in Case 3, the sample size is larger than 30 and therefore the t-distribution can be approximated by the z-distribution.

Step 3 Signficance level α is normally specified by the decision-maker and is usually between .01 and .1.

Step 4 Method A

The critical value(s) and the decision rule in terms of the value of the test statistic:

Alternative	***Critical value***	***Decision rule***
Case 1 ***When σ^2 is known***		
$H_A: \mu \neq \mu_o$	$\pm z_{\alpha/2}$	Reject H_o if $\lvert z_{calculated} \rvert > z_{\alpha/2}$
$H_A: \mu < \mu_o$	$-z_\alpha$	Reject H_o if $z_{calculated} < -z_\alpha$
$H_A: \mu > \mu_o$	$+z_\alpha$	Reject H_o if $z_{calculated} > z_\alpha$
Case 2 ***When σ^2 is unknown***		
$H_A: \mu \neq \mu_o$	$\pm t_{\alpha/2,n-1}$	Reject H_o if $\lvert t_{calculated} \rvert > t_{\alpha/2,n-1}$
$H_A: \mu < \mu_o$	$-t_{\alpha,n-1}$	Reject H_o if $t_{calculated} < -t_{\alpha,n-1}$
$H_A: \mu > \mu_o$	$+t_{\alpha,n-1}$	Reject H_o if $t_{calculated} > t_{\alpha,n-1}$
Case 3 ***When σ^2 is unknown and $n > 30$***		
$H_A: \mu \neq \mu_o$	$\pm z_{\alpha/2}$	Reject H_o if $\lvert z_{calculated} \rvert > z_{\alpha/2}$
$H_A: \mu < \mu_o$	$-z_\alpha$	Reject H_o if $z_{calculated} < -z_\alpha$
$H_A: \mu > \mu_o$	$+z_\alpha$	Reject H_o if $z_{calculated} > z_\alpha$

Method B

The decision rule in terms of the p-value:

Reject H_o if p-value $< \alpha$.

where the *p*-value is the minimum level of significance required to reject the null hypothesis H_o.

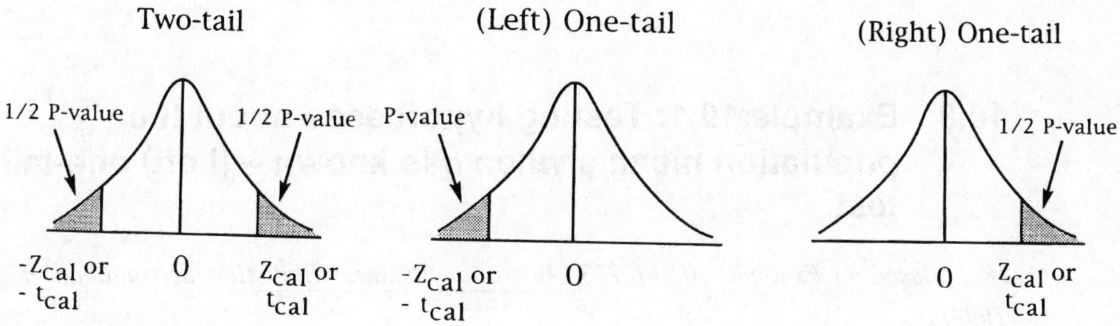

<u>Step 5</u> Method A

The calculated sample (observed) value of the test statistic under the null hypothesis:

<u>Case 1</u> ***<u>When σ^2 is known</u>*** $z_{calculated} = \dfrac{\overline{X} - \mu_0}{\sigma/\sqrt{n}}$

<u>Case 2</u> ***<u>When σ^2 is unknown</u>*** $t_{calculated} = \dfrac{\overline{X} - \mu_0}{s/\sqrt{n}}$

<u>Case 3</u> ***<u>When σ^2 is unknown and n > 30</u>*** $Z_{calculated} = \dfrac{\overline{X} - \mu_0}{s/\sqrt{n}}$

<u>Method B</u>

Calculating the p-value:

<u>Case 1</u> ***<u>When σ^2 is known</u>*** $z_{calculated} = \dfrac{\overline{X} - \mu_0}{\sigma/\sqrt{n}}$

If H_A: $\mu \neq \mu_o$ *p*-value $= 2\,P[Z > ABS(z_{calculated})]$

If H_A: $\mu < \mu_o$ *p*-value $= P[Z < z_{calculated}]$

If H_A: $\mu > \mu_o$ *p*-value $= P[Z > z_{calculated}]$

<u>Case 2</u> ***<u>When σ^2 is unknown</u>*** $t_{calculated} = \dfrac{\overline{X} - \mu_0}{s/\sqrt{n}}$

If H_A: $\mu \neq \mu_o$ *p*-value $= 2\,P[t > ABS(t_{calculated})]$

If H_A: $\mu < \mu_o$ *p*-value $= P[t < t_{calculated}]$

If H_A: $\mu > \mu_o$ *p*-value $= P[t > t_{calculated}]$

<u>Case 3</u> ***<u>When σ^2 is unknown and n > 30</u>*** $Z_{calculated} = \dfrac{\overline{X} - \mu_0}{s/\sqrt{n}}$

If H_A: $\mu \neq \mu_o$ *p*-value $= 2\,P[Z > ABS(z_{calculated})]$

If H_A: $\mu < \mu_o$ *p*-value $= P[Z < z_{calculated}]$

If H_A: $\mu > \mu_o$ *p*-value $= P[Z > z_{calculated}]$

Step 6 Conclusion: Compare the observed value of the test statistic (Method A) or the *p*-value (Method B) calculated in Step 5 with the decision rule in Step 4, make a decision and answer the question.

10.3 Example 10.1: Testing hypotheses about the population mean μ when σ is known – (Left) one-tail test

(Ref: Based on Example 10.3, p.273, Australian Business Statistics, Selvanathan et al., 1994)

The manager of a department store is thinking about establishing a new billing system for the store's credit customers. After a thorough financial analysis, she determines that the new system will not be cost effective if the average monthly account is less than \$70. A random sample of 200 monthly accounts is drawn, for which the mean monthly account is \$66. With $\alpha = .05$, is there sufficient evidence to conclude that the new system will not be cost effective? Assume that the population standard deviation is \$30. Perform the following tasks as intermediate steps to reach your conclusion. Use EXCEL to perform Tasks 4 and 5. Include in your EXCEL file information regarding the other tasks.

Task 1 Determine the null and alternative hypotheses.
Task 2 Determine the test statistic and its sampling distribution.
Task 3 Specify the level of significance.
Task 4 Obtain the critical value(s) and develop the decision rule.
Task 5 Obtain the sample value of the test statistic and the corresponding *p*-value.
Task 6 Derive your conclusion for the test in two ways:
- (a) using the sample value of the test statistic obtained in Task 5 and the decision rule developed in Task 4.
- (b) using the *p*-value obtained in Task 5 and the level of significance given in Task 3.

Task 7 Save your workbook as 'A:EX101.XLS' on your disk and print the file.

Solution to Example 10.1

Let X be the amount of a monthly account, μ be the mean amount of all monthly accounts and σ be the population standard deviation. In cell A1, type the title 'Hypothesis Testing about Population Mean Mu' and center it across cells A1 to C1. Type the second title 'Population Standard Deviation Sigma is Known' in cell A2 and center it across cells A2 to C2.

Data summary (given)
Sample size n = 200; sample mean $\overline{X}$ = \$66; population standard deviation σ (known) = \$30; and significance level α = .05. In cells B4 to B6, type the labels 'Sample size n =',

'Sample mean Xbar =' and 'Population SD sigma ='. To input the given information, in cells C4 to C6, type the corresponding values **200, 66** and **30.**

Task 1 The new system will not be cost effective if the average monthly account is less than \$70. Therefore, the null and the alternative hypotheses for this test are

$$H_0: \quad \mu = 70$$
$$H_A: \quad \mu < 70$$

As the alternative hypothesis contains the 'less than' sign, the rejection region of the test is in the left hand tail and this is a one-tail test.

In cells B8 and B9, type the labels 'Ho: Mu =' and 'HA: Mu < ' and, in cells C8 and C9, type the values **70** and **70.**

Task 2 As the sample size n=200>30, using the Central Limit Theorem, $\overline{X}$ is distributed approximately normal. Therefore, when the population standard deviation σ is *known*, to test a hypothesis about the population mean μ, we use the test statistic (see Case 1, Section 10.2)

$$Z = \frac{\overline{X} - \mu}{\sigma/\sqrt{n}}$$

which is distributed as the standard normal distribution as $\overline{X}$ is normal and σ is known. In cells B11 and C11, type the labels 'Test Statistic:' and 'Z = (Xbar – mu)/(sigma/sqrt(n))'.

Task 3 In cell B13, type the label 'Level of significance =' and, in cell C13, type the given value 0.05.

Task 4 ***Calculating the critical value(s)***

As this is a *(left) one-tail* test (see diagram below), for a test with 5 percent level of significance,

$$\text{Critical value } z_{critical} = -z_{\alpha} = -z_{.05}.$$

This value is based on a standard normal distribution and can be obtained using the EXCEL function **NORMSINV**. In cell B15, type the label 'Critical value z(.05) ='. Click on cell C15 to store the result. In this example, since H_A: μ < 70, the critical region is on the left tail of the z-distribution with area .05. To obtain this value, click on **Paste Function** and then select *Functional Category* **Statistical**. From the *Function Name* select **NORMSINV** [this function will return the z-value associated with a given cumulative probability] and click **OK** on the dialogue box. The NORMSINV dialogue box will appear on the screen [see Figure 10.1].

Figure 10.1 NORMSINV Function dialogue box

This will ask for one input, the (cumulative) *probability*. In this example, as the critical region is on the left tail, the cumulative probability is the same as the level of significance $\alpha = .05$. [It should be noted that if the critical region is on the right tail, then the cumulative probability would be 1- level of significance = $1 - \alpha = .95$.] After typing this value **.05** or the corresponding cell reference **C13**, click **OK**. You will see the $z_{critical}$ value -1.644853 appearing in the selected cell C15.

Alternatively, you can type the formula **=NORMSINV(.05)** or **=NORMSINV(C13)** in cell C15 to obtain this value.

The decision rule

Therefore the decision rule for the test is

Method A: **Reject H_0 if $z_{calculated} < z_{critical} = -1.644853$**

Method B: **Reject Ho if *p*-value < $\alpha = .05$**

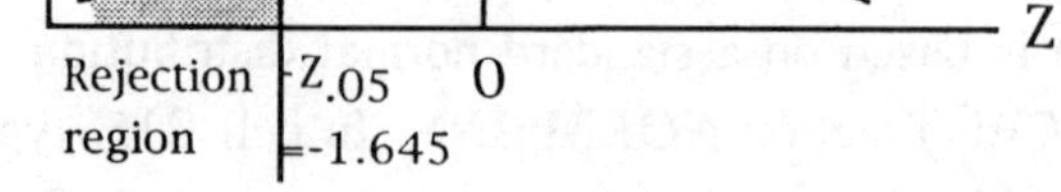

In cells B16 to B18, type the labels 'Decision Rule: ', 'Method A: Reject Ho if Zcal <' and 'Method B: Reject Ho if p-value <'. In cells C17 and C18, type the formulas **=C15** and **=C13**.

Task 5 ***Calculating the data-based value of the test statistic ($z_{calculated}$)***

In cells B20 to B21, type the labels 'SE of sample mean =' and 'Test statistic Z-calculated ='. To obtain the standard error of the sample mean, in cell C20, type the formula **=C6/SQRT(C4)**.

Select cell C21 to store the sample value of the test statistic.

Method 1 Given $\overline{X} = 66$, $\sigma = 30$, $n = 200$ and $\mu = 70$ (under H_0). To calculate the data-based value of the test statistic, in cell C21, type the formula **=(C5-C8)/C20**. This will give the $z_{calculated}$ value (= -1.88562) of the test statistic.

Method 2 You can also use the **Paste Function** to obtain the $z_{calculated}$ value. Click on cell C21 to store the normalised value. First click on fx **Paste Function** and then select the *Function Category* **Statistical**. From the *Function Name* select **STANDARDIZE** and then click **OK** on the dialogue box. Now you will see a *Standardize function* dialogue box to appear (See Figure 10.2).

Figure 10.2 STANDARDIZE Function dialogue box

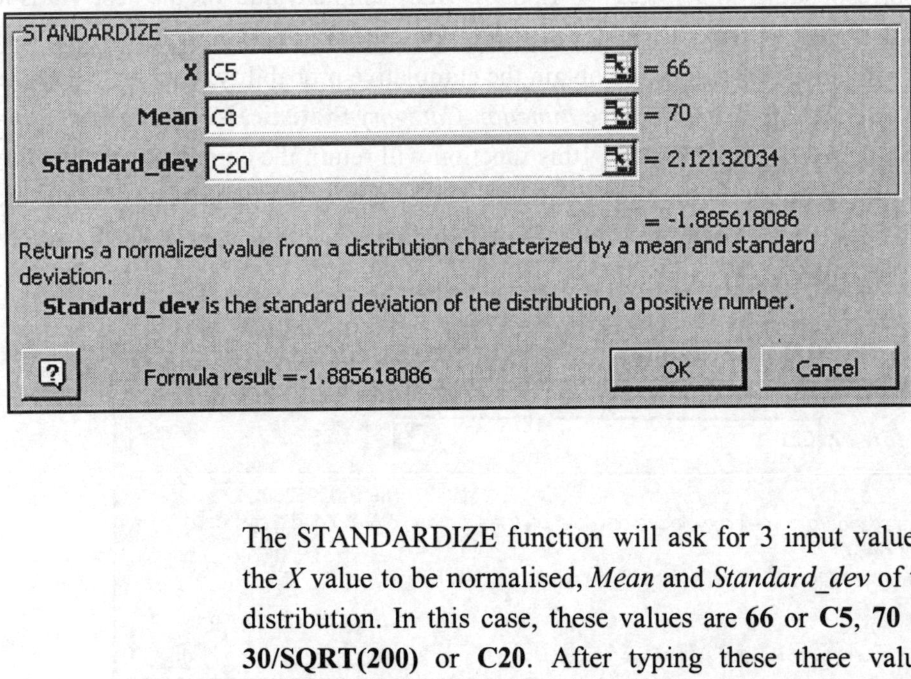

The STANDARDIZE function will ask for 3 input values, namely, the *X* value to be normalised, *Mean* and *Standard_dev* of the normal distribution. In this case, these values are **66** or **C5**, **70** or **C8** and **30/SQRT(200)** or **C20**. After typing these three values or the corresponding cell references, click **OK.** You will see the standardised value $z_{calculated} = -1.88562$ appearing in cell C21.

Method 3 Instead of going through the fx **Paste Function**, to obtain the z-value, type the formula **=STANDARDIZE(66,70,30/SQRT(200))** or **=STANDARDIZE(C5,C8,C20)** in cell C21.

Calculating the p-value

For a left-tail z-test, the *p*-value = P(Z < $z_{calculated}$), where $z_{calculated}$ is the standardised value of the test statistic calculated above.

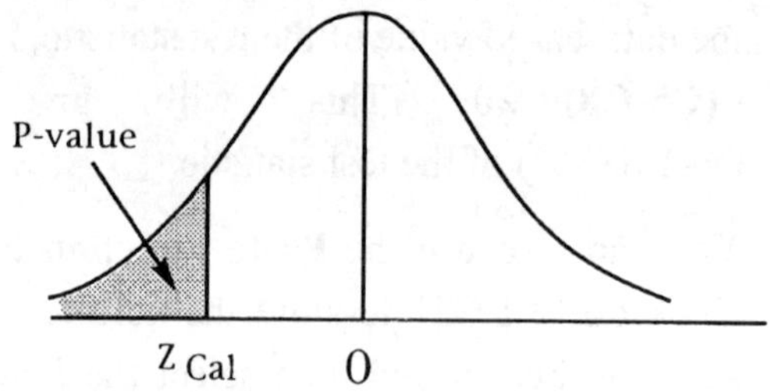

To calculate this probability, first type the label 'Cumulative probability =' in cell B22 and select cell C22 to store the cumulative probability corresponding to $z_{calculated}$ = -1.88562 (the sample value of the test statistic calculated above). Thus, the *p*-value for a (left) one-tail test will be this cumulative probability. To obtain the cumulative probability, click on *fx* **Paste Function** and then select the *Function Category* **Statistical**. From the *Function Name* select **NORMSDIST** [this function will return the cumulative probability of the standard normal distribution for a given z-value] and click **OK** on the dialogue box. Now, the NORMSDIST dialogue box will appear on the screen (see Figure 10.3).

Figure 10.3 NORMSDIST Function dialogue box

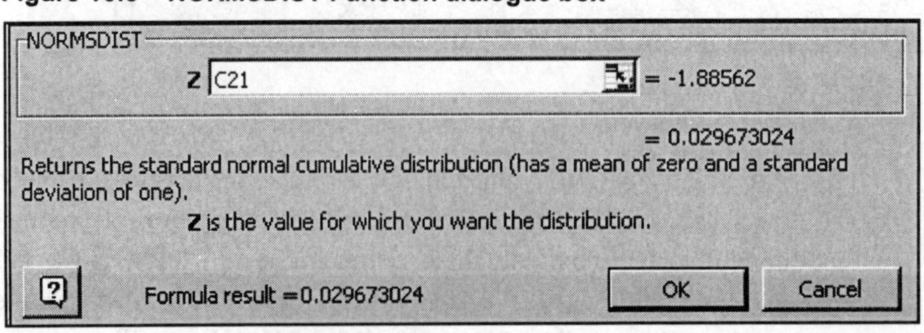

The **NORMSDIST** function will request one input, namely, *Z*, the standardised value for which the cumulative probability is sought, in this case the input value *z* is -1.88562. After typing this value **-1.88562** or its cell reference **C21**, on the dialogue box, click **OK**. The cumulative probability will appear in cell C22 as .029673. In cell B23, type the label 'p-value ='. In cell C23, type the formula **=C22**. The value .029673 will appear in cell C23.

Alternatively, without going through the *Paste Function*, you can also straight-away type the formula to obtain the *p*-value. First, in cell B23, type 'p-value ='. Then use the formula for the standard normal cumulative probability corresponding to the standardised value of the test statistics. That is, in cell C23, type the formula **=NORMSDIST(-1.88562)** or **=NORMSDIST(C21)** or **NORMDIST(C21,0,1,TRUE)**. This will give the required *p*-value for a (left) one-tail test.

Task 6 ***Conclusion***

(a) Method A From Task 5, $z_{calculated} = -1.88562$

From Task 4, $z_{critical} = -1.644853$

Based on the decision rule developed in Task 4, as $z_{calculated} = -1.88562 < z_{critical} = -1.644853$, we reject H_O: $\mu = 70$ in favour of the alternative H_A: $\mu < 70$ (see the diagram below). In cells B25 and C25, type the text, 'Based on Method A' and 'Since Zcal = -1.88 < Zcrit = -1.645, reject Ho'.

(b) Method B From Task 5, *p*-value = .029673

Based on the decision rule given in Task 4, as the level of significance $\alpha = .05 >$ *p*-value, we reject H_O: $\mu = 70$ in favour of the alternative H_A: $\mu < 70$ (see the diagram below). In cells B26 and C26, type the text, 'Based on Method B' and 'Since p-value = .0297 < alpha = .05, reject Ho'.

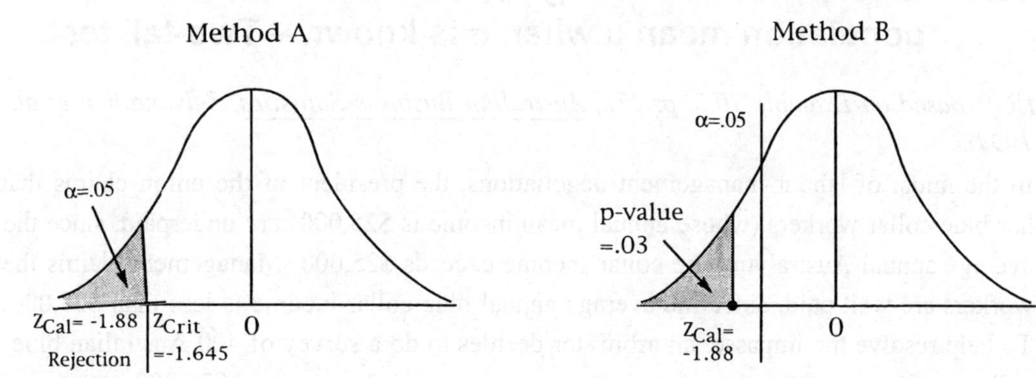

Task 7 Save your workbook as **A:EX101.XLS** on your disk and print the file. The output from Tasks 1-6 is given in Figure 10.4.

Figure 10.4 Hypothesis testing about μ when σ is known: Left one-tail test

	A	B	C	
1		Hypothesis Testing about Population Mean Mu		
2		Population Standard Deviation Sigma is Known		
3				
4	Data Summary:	Sample size n =		200
5		Sample mean Xbar =		66
6		Population SD sigma =		30
7				
8	Step 1	Ho: Mu =		70
9		HA :Mu <		70
10				
11	Step 2	Test statistic	Z=(Xbar - mu)/(sigma/sqrt(n))	
12				
13	Step 3	Level of significance =		0.05
14				
15	Step 4	Critical value z(.05) =		-1.644853
16		Decision rule:		
17		Method A: Reject Ho if Zcal <		-1.644853
18		Method B: Reject Ho if p-value <		0.05
19				
20	Step 5	SE of sample mean =		2.121320344
21		Test statistic Z-calculated =		-1.885618083
22		Cumulative probability =		0.029673154
23		p-value =		0.029673154
24				
25	Step 6	Based on Method A	Since Zcal = -1.88 < Zcrit = -1.645, reject Ho	
26		Based on Method B	Since p-val = .0297 < alpha = .05, reject Ho	

10.4 Example 10.2: Testing hypotheses about the population mean μ when σ is known – Two-tail test

(Ref: Based on Example 10.2, p. 272, Australian Business Statistics, Selvanathan et al., 1994)

In the midst of labour-management negotiations, the president of the union claims that her blue-collar workers (whose annual mean income is \$25,000) are underpaid, since the average annual Australian blue-collar income exceeds \$25,000. Management claims the workers are well paid, since the average annual blue-collar income is less than \$25,000. To help resolve the impasse, an arbitrator decides to do a survey of 400 Australian blue-collar workers to determine if their mean income is different from \$25,000. Assuming that σ = \$5000, can the arbitrator conclude at the 5 percent significance level that μ is different from \$25,000 if the sample mean is $\overline{X}$=\$25,312.5. Perform the following tasks as intermediate steps to reach your conclusion.

Tasks 1-6 Perform the same Tasks 1-6 as in the Example 10.1 question. Use EXCEL to perform Tasks 4 and 5. Include in your EXCEL file information regarding the other tasks.

Task 7 Save your workbook as 'A:EX102.XLS' on your disk and print the file.

Solution to Example 10.2

Let X be the annual income of an Australian blue-collar worker, μ be the mean annual income and σ be the population standard deviation. In cell A1, type the title 'Hypothesis Testing about Population Mean Mu' and center it across columns A to D. Type the second title 'Population Standard Deviation Sigma is Known' in cell A2 and center it across columns A to D.

Data summary (given)

Sample size n = 400; sample mean $\overline{X}$ = \$25,312.5; population standard deviation σ (known) = \$5000; significance level α = .05. In cells B4 to B6, type the labels 'Sample size n =', 'Sample mean Xbar =' and 'Population SD sigma ='. To input the given information, in cells C4 to C6, type the corresponding values **400**, **25312.5** and **5000**.

Task 1 The arbitrator wants to determine if the Australian blue-collar workers mean income is different from \$25,000. Therefore, the null and the alternative hypotheses are

$$H_0: \quad \mu = 25000$$
$$H_A: \quad \mu \neq 25000$$

As the alternative hypothesis contains the 'not equal sign', i.e. the rejection region is in both tails, this is a two-tail test.

In cells B8 and B9, type the labels 'Ho: Mu ='and 'HA: Mu not equal to ' and, in cells C8 and C9, type the values **25000** and **25000**.

Task 2 As the sample size n=400>30, using the Central Limit Theorem, $\overline{X}$ is distributed approximately normal. Therefore, when the population standard deviation σ is <u>known</u>, to test a hypothesis about the population mean μ, we use the test statistic (see <u>Case 1</u>, Section 10.2)

$$Z = \frac{\overline{X} - \mu}{\sigma/\sqrt{n}}$$

which is distributed as the standard normal distribution as $\overline{X}$ is normal and σ is known. In cells B11 and C11, type the labels 'Test Statistic:' and 'Z = (Xbar – mu)/(sigma/sqrt(n))'.

Task 3 In cell B13, type the label 'Level of significance =' and, in cell C13, type the given value 0.05.

Task 4 ***<u>Calculating the critical value(s)</u>***

As this is a *two-tail test*, for a test with 5 percent level of significance, we have two critical regions one on each tail of the z-distribution with tail areas of 0.025. Therefore,

$$\text{Critical values } z_{critical} = \pm z_{\alpha/2} = -z_{.025} \text{ and } +z_{.025}.$$

These values are based on a standard normal distribution (see diagram below) and can be obtained using the EXCEL function **NORMSINV**. In cell B15, type the label 'Critical values -z(.025) and +z(.025) ='. Click on cell C15 to store the result for $-z_{.025}$. Using the same procedure as in Task 4 of Example 10.1, type the formula **=NORMSINV(.025)** or **=NORMSINV(C13/2)** in cell C15 to obtain this value. Click on cell D15 to store the value for $+z_{.025}$ which is the negative of $-z_{.025}$. In cell D15, type the formula **=-C15**. You will see the two $z_{critical}$ values -1.959961 and 1.959961 appearing in cells C15 and D15, respectively.

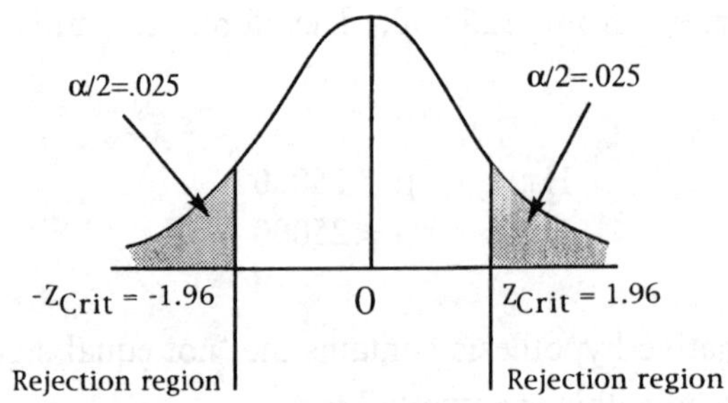

The decision rule

Therefore the decision rule for the test is (see diagram below)

Method A: **Reject H_0 if $|z_{calculated}| > z_{critical} = 1.959961$**

That is, Reject H_0 if $z_{calculated} < -z_{critical} = -1.959961$

or $z_{calculated} > z_{critical} = 1.959961$

Method B: **Reject Ho if *p*-value < α = .05.**

In cells B16 to B18, type the labels 'Decision Rule: ', 'Method A: Reject Ho if l(Zcal)l >' and 'Method B: Reject Ho if p-value <'. In cells C17 and C18, type the formulas **=C15** and **=C13**.

Task 5 ***Calculating the data-based value of the test statistic ($z_{calculated}$)***

In cells B20 to B21, type the labels 'SE of sample mean =' and 'Test statistic Z-calculated ='. To obtain the standard error of the sample mean, in cell C20, type the formula **=C6/SQRT(C4)**.

Select cell C21 to store the sample value of the test statistic. Given $\overline{X}$ = 25312.5, σ = 5000, n = 400 and μ = 2500 (under H_0). To calculate the data-based value of the test statistic, in cell C21, type the formula **=(C5-C8)/C20**. This will give the $z_{calculated}$ value (= 1.25) of the test statistic.

Alternatively, you can type the formula **=STANDARDIZE (25312.5,25000,5000/SQRT(400))** or **=STANDARDISE(C5,C8,C20)** in cell C21.

Calculating the p-value

For a two-tail z-test, the *p*-value = 2P[Z > ABS($z_{calculated}$)], where $z_{calculated}$ is the standardised value of the test statistic calculated above (see the diagram below).

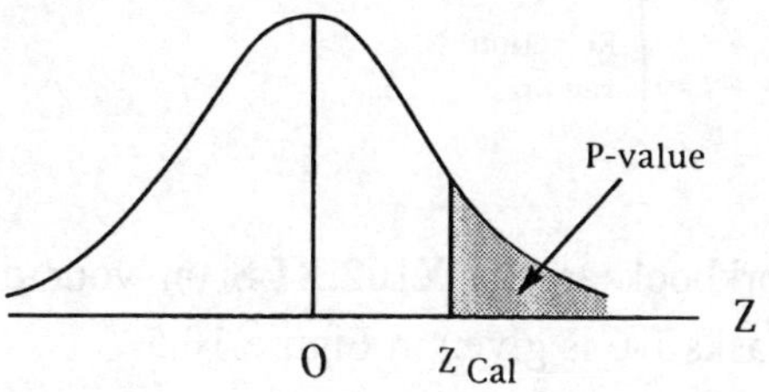

In cell B22 and B23, type the labels 'Cumulative probability =' and 'p-value=' and click the mouse on C22. Now, we use the function **NORMSDIST** or **NORMDIST** to obtain the cumulative probability corresponding to the absolute value of $z_{calculated}$ = 1.25 (from above) and then obtain the *p*-value as 2(1-cumulative probability). That is, in cell C22, type the formula **=NORMSDIST(ABS(C21))** or **=NORMDIST(ABS(C21),0,1,TRUE)**. To calculate the *p*-value, in cell C23, type the formula **=2*(1-C22).**

Task 6 ***Conclusion***

(a) Method A From Task 5, $z_{calculated}$ = 1.25

From Task 4, $z_{critical}$ = -1.96 and 1.96

Based on the decision rule developed in Task 4, as -1.96 < $z_{calculated}$ = 1.25 < 1.96, we do not reject H_O: μ = 25000 in favour of the alternative H_A: μ ≠ 25000 (see the diagram below). In cells B25 and C25, type the text, 'Based on Method A' and 'Since -1.96 < Zcal = 1.25 < Zcrit = 1.96, do not reject Ho'.

(b) Method B From Task 5, p-value = .2113

Based on the decision rule given in Task 4, as the level of significance α = .05 < *p*-value, we do not reject H_O: μ = 25000 in favour of the alternative H_A: μ ≠ 25000 (see the diagram below). In cells B26 and C26, type the text, 'Based on Method B' and 'Since p-value = .2113 > alpha = .05, do not reject Ho'.

Method A

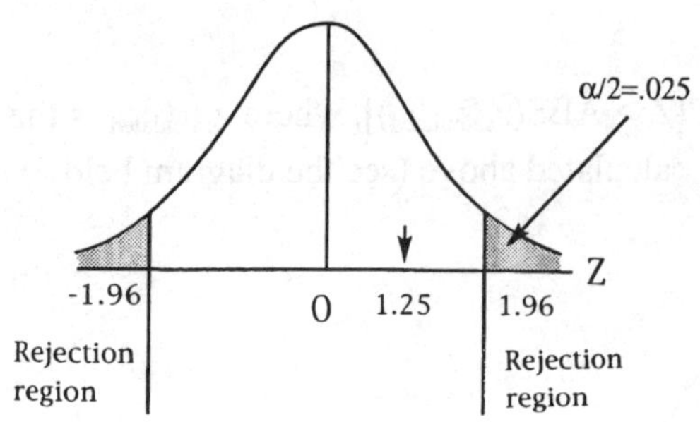

Method B

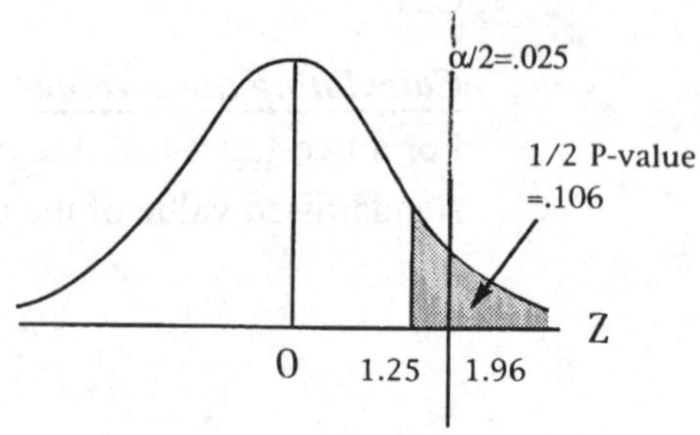

Task 7 Save your workbook as **A:EX102.XLS** on your disk and print the file. The output from Tasks 1-6 is given in Figure 10.5.

Figure 10.5 Hypothesis testing about μ when σ is known: Two-tail test

	A	B	C	D
1		**Hypothesis Testing about Population Mean Mu**		
2		**Population Standard Deviation Sigma is Known**		
3				
4	Data Summary:	Sample size n =	400	
5		Sample mean Xbar =	25312.5	
6		Population SD sigma =	5000	
7				
8	Step 1	Ho: Mu =	25000	
9		H1 :Mu not equal to	25000	
10				
11	Step 2	Test statistic	Z=(Xbar - mu)/(sigma/sqrt(n))	
12				
13	Step 3	Level of significance =	0.05	
14				
15	Step 4	Critical values -z(.025) and +Z(.025)=	-1.959961082	1.95996
16		Decision rule:		
17		Method A: Reject Ho if I(Zcal)I >	1.959961082	
18		Method B: Reject Ho if p-value <	0.05	
19				
20	Step 5	SE of sample mean =	250	
21		Test statistic Z-calculated =	1.25	
22		Cumulative probability =	0.894350161	
23		p-value =	0.211299678	
24				
25	Step 6	Based on Method A	Since -1.96 < Zcal = 1.25 < Zcrit = 1.96, do not reject Ho	
26		Based on Method B	Since p-value = .21 > alpha = .05, do not reject Ho	

10.5 Example 10.3: Testing hypotheses about the population mean μ when σ is known – (Right) one-tail test

A fast food chain operator is investigating the annual food expenditure patterns of families in a country town in Warwick in view of opening a new outlet there. The following data are the annual food expenditure (in $'000) of 10 families randomly selected from that country town.

4, 8, 12, 11, 14, 6, 12, 8, 9, 5

Using the sample data, test at the 10% level of significance whether the average annual food expenditure of the country town is more than seven thousand dollars; i.e. H_A: $\mu > 7$, by answering the following questions. (Assume that all annual family food expenditure from the country town follows a normal distribution with standard deviation 3.) Perform the following tasks as intermediate steps to reach your conclusion.

Tasks 1-6 Perform the same Tasks 1-6 as in the Example 10.1 question. Use EXCEL to perform Tasks 4 and 5. Include in your EXCEL file information regarding the other tasks.

Task 7 Save your workbook as 'A:EX103.XLS' on your disk and print the file.

Solution to Example 10.3

One major difference between this example and the last two examples is that, in Examples 10.1 and 10.2, the raw data were not given but the sample mean values were given. In this example, we have access to the raw data. Therefore, we can calculate the value of the sample mean.

Let X be the annual expenditure on food (in '000 dollars) of a family in Warwick, μ be the mean expenditure and σ be the population standard deviation. In cell A1, type the title 'Hypothesis Testing about Population Mean Mu' and center it across cells A1 to D1. Type the second title 'Population Standard Deviation is Known' in cell A2 and center it across cells A2 to D2.

Data summary (given)

Sample size n = 10; sample mean $\overline{X}$ = to be calculated; population standard deviation σ (known) = $3; significance level α = .10; population variable X, the annual expenditure on food (in '000 dollars) of a family in Warwick, is normally distributed. In cells B4 to B6, type the labels 'Sample size n =', 'Sample mean Xbar =' and 'Population SD sigma ='. To input the given information, in cells C4 and C6, type the corresponding values **10** and **3.** In cell D4, type the column heading 'Data' and type the given 10 family food expenditures in cells D5 to D14. To obtain the sample mean $\overline{X}$ from the data set, in cell C5, type the formula **=AVERAGE(D5:D14).**

Task 1 The null and the alternative hypotheses are

$$H_0: \quad \mu = 7$$
$$H_A: \quad \mu > 7$$

As the alternative hypothesis contains the 'greater than' sign, the rejection region of the test is in the right hand tail and this is a one-tail test.

In cells B8 and B9, type the labels 'Ho: Mu =' and 'HA: Mu > ' and, in cells C8 and C9, type the values **7** and **7**.

Task 2 As the population variable X is normally distributed, $\overline{X}$ is distributed exactly normal. Therefore, when the population standard deviation σ is <u>*known*</u>, to test a hypothesis about the population mean μ, we use the test statistic (see <u>Case 1</u>, Section 10.2)

$$Z = \frac{\overline{X} - \mu}{\sigma/\sqrt{n}}$$

which is distributed as the standard normal distribution as $\overline{X}$ is normal and σ is known.

In cells B11 and C11, type the labels 'Test Statistic:' and 'Z = (Xbar – mu)/(sigma/sqrt(n))'.

Task 3 In cell B13, type the label 'Level of significance =' and in cell C13, type the given value **0.10**.

Task 4 <u>***Calculating the critical value(s)***</u>

As this is a *(right) one-tail* test (see diagram below), for a test with 10 percent level of significance,

$$\text{Critical value } z_{critical} = z_{\alpha} = z_{.10}$$

This value is based on a standard normal distribution and can be obtained using the EXCEL function NORMSINV. In cell B15, type 'Critical value z(.10)='. As before in Task 4 of Example 10.1, to obtain the critical value, in cell C15, type the formula **=NORMSINV(.90)** or **=NORMSINV(1-C13)**. The critical value will be 1.281551.

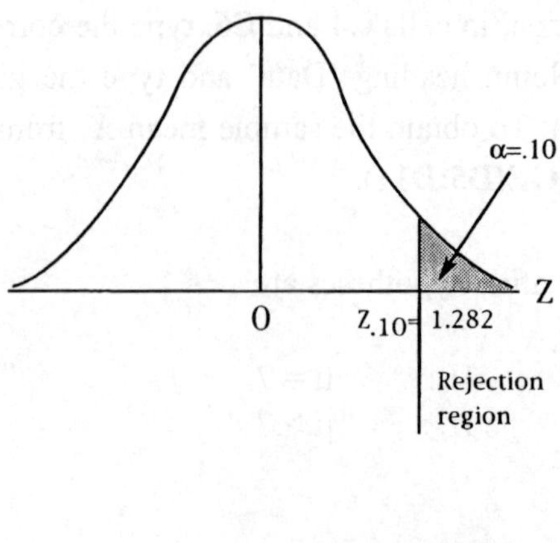

The decision rule

Therefore the decision rule for the test is

Method A: **Reject H_0 if $z_{calculated} > z_{critical} = 1.281551$**
Method B: **Reject Ho if *p*-value < $\alpha = .10$**

In cells B16 to B18, type the labels 'Decision Rule: ', 'Method A: Reject Ho if Zcal >' and 'Method B: Reject Ho if p-value <'. In cells C17 and C18, type the cell references **=C15** and **=C13**.

Task 5 ***Calculating the data-based value of the test statistic ($z_{calculated}$)***

In cells B20 and B21, type the labels 'SE of sample mean =', 'Test statistic Z-calculated ='. To obtain the standard error of the sample mean, in cell C20, type the formula **=C6/SQRT(C4)**.

Select cell C21 to store the sample value of the test statistic. To calculate the data-based value of the test statistic, in cell C21, type the formula **=(C5-C8)/C20**. Now this will give the value (=2.002776) for the test statistic.

Calculating the p-value

For a right-tail z-test, the *p*-value = $P(Z > z_{calculated})$, where $z_{calculated}$ is the standardised value of the test statistic.

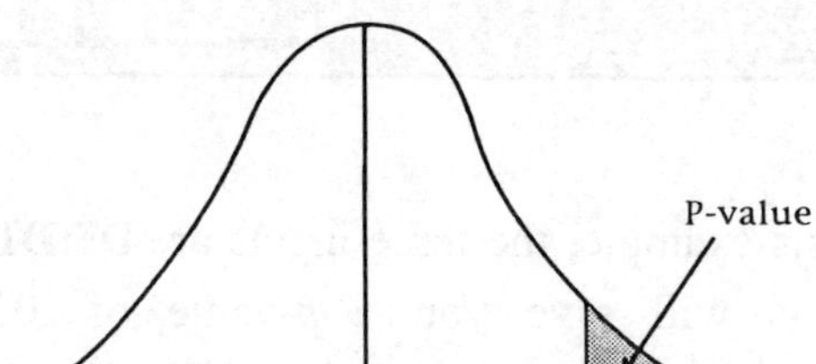

Method 1 To calculate this probability, first type the label 'Cumulative probability =' in cell B22 and select cell C22 to store the cumulative probability corresponding to $z_{calculated} = 2.002776$ (the sample value of the test statistic calculated above). Thus, the *p*-value for a (right) one-tail test will be one minus this cumulative probability. Now we use the function **NORMSDIST** or **NORMDIST** to obtain the cumulative probability corresponding to the z-value = 2.002776. Then we calculate the *p*-value as (1-cumulative probability). Since H_A: $\mu > 7$, we use the right tail. That is, in cell C22, type the formula **=NORMSDIST(2.002776)** or **=NORMSDIST(C21)** or **=NORMDIST(2.002776,0,1,TRUE)** or =

NORMDIST(C21,0,1,TRUE). To calculate the *p*-value, in cell B23, type the label 'p-value=' and, in cell C23, type the formula **=1-C22**.

Method 2 In this case, as the raw data are given, we can use the *Function Name* **ZTEST** under the *Function Category* **Statistical** of the *f*x **Paste Function** to automatically obtain the *p*-value. In cell B23, type the label 'p-value=' and click the mouse on C23. When you select **ZTEST** and click **OK**, the dialogue box for **ZTEST** appears on the screen and will ask for 3 inputs (see Figure 10.6), namely, *Array* (the raw data), *X* (the value to be tested under H_0) and *Sigma* (the population standard deviation).

Figure 10.6 ZTEST Function dialogue box

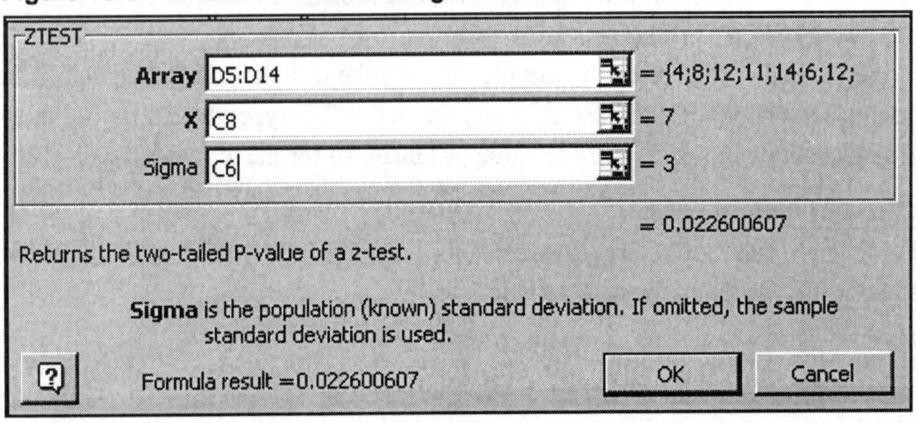

In this example, the three inputs are **D5:D14**; **7** or **C8**; and **3** or **C6**. This will give you a *p*-value of .022601 in cell C23. *Alternatively*, type the formula **=ZTEST(D5:D14,C8,C6)** in cell C23.

Task 6 ***Conclusion***

(a) Method A From Task 5, $z_{calculated} = 2.002776$

From Task 4, $z_{critical} = 1.281551$

Based on the decision rule developed in Task 4, as $z_{calculated} = 2.002776 > z_{critical} = 1.281551$, we reject H_O: $\mu = 7$ in favour of the alternative H_A: $\mu > 7$ (see the diagram below). In cells B25 and C25, type the text, 'Based on Method A' and 'Since Zcal = 2.00 > Zcrit = 1.282, reject Ho'.

(b) Method B From Task 5, *p*-value = .0226

Based on the decision rule given in Task 4, as the level of significance $\alpha = .10 >$ *p*-value, we reject H_O: $\mu = 7$ in favour of the alternative H_A: $\mu > 7$ (see the diagram below). In cells B26 and C26, type the text, 'Based on Method B' and 'Since p-value = .0226 < alpha = .10, reject Ho'.

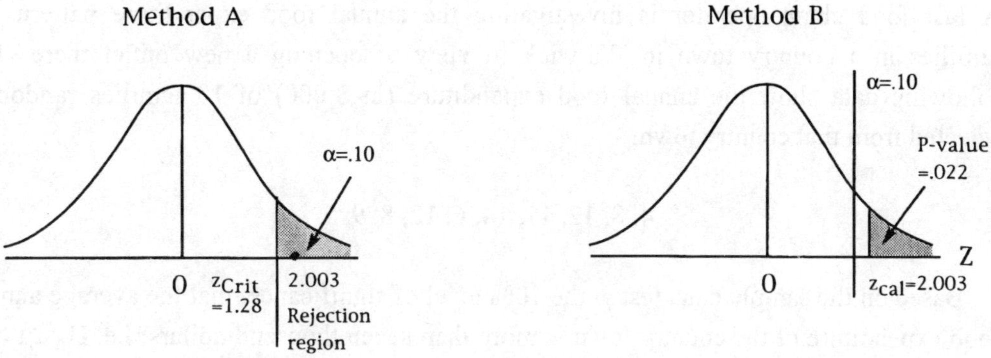

Task 7 Save your workbook as **A:EX103.XLS** on your disk and print the file. The output from Tasks 1-6 is given in Figure 10.7.

Figure 10.7 Hypothesis testing about μ when σ is known: Right one-tail test

	A	B	C	D
1		Hypothesis Testing about Population Mean Mu		
2		Population Standard Deviation is Known		
3				
4	Data Summary:	Sample size n =	10	Data
5		Sample mean Xbar =	8.9	4
6		Population SD sigma =	3	8
7				12
8	Step 1	Ho: Mu =	7	11
9		HA: Mu >	7	14
10				6
11	Step 2	Test statistic	Z=(Xbar - mu)/(sigma/sqrt(n))	12
12				8
13	Step 3	Level of significance =	0.1	9
14				5
15	Step 4	Critical value z(.10) =	1.281550794	
16		Decision rule:		
17		Method A: Reject Ho if Zcal >	1.281550794	
18		Method B: Reject Ho if p-value <	0.1	
19				
20	Step 5	SE of sample mean =	0.948683298	
21		Test statistic Z-calculated =	2.002775851	
22		Cumulative probability =	0.977399393	
23		p-value =	0.022600607	
24				
25	Step 6	Based on Method A	Since Zcal = 2.00 > Zcrit = 1.28, reject Ho	
26		Based on Method B	Since p-val = .022 < alpha = .10, reject Ho	

10.6 Example 10.4: Testing hypotheses about the population mean μ when σ is unknown

A fast food chain operator is investigating the annual food expenditure patterns of families in a country town in Warwick in view of opening a new outlet there. The following data show the annual food expenditure (in \$'000) of 10 families randomly selected from that country town.

4, 8, 12, 11, 14, 6, 12, 8, 9, 5

Based on the sample data, test at the 10% level of significance, that the average annual food expenditure of the country town is more than seven thousand dollars; i.e. $H_A : \mu > 7$ by answering the following questions. (Assume that all annual family food expenditure from that country town follow a normal distribution.) Perform the following tasks as intermediate steps to reach your conclusion.

Tasks 1-6 Perform the same Tasks 1-6 as in the Example 10.1 question. Use EXCEL to perform Tasks 4 and 5. Include in your EXCEL file information regarding the other tasks.

Task 7 Save your workbook as 'A:EX104.XLS' on your disk and print the file.

Solution to Example 10.4

One major difference between this example and the last three examples is that, in Examples 10.1, 10.2 and 10.3, the population standard deviation σ was given. In this example, the population standard deviation σ is unknown and we need to estimate it by the sample standard deviation. Since, in this example, we have access to the raw data, we can calculate the value of the sample standard deviation. Example 10.3 and this example are exactly the same except for the standard deviation.

Let X be the annual expenditure on food (in '000 dollars) of a family in Warwick, μ be the mean expenditure σ be the population standard deviation. In cell A1, type the title 'Hypothesis Testing about Population Mean Mu' and center it across cells A1 to D1. Type the second title 'Population Standard Deviation Sigma is Unknown' in cell A2 and center it across cells A2 to D2.

<u>Data summary (given)</u>

Sample size $n = 10$; sample mean $\overline{X}$ = to be calculated; population standard deviation σ (unknown) to be estimated by s = to be calculated; significance level $\alpha = .10$; population variable X, annual expenditure on food (in '000 dollars) of a family in Warwick, is normally distributed. In cells B4 to B6, type the labels 'Sample size n =', 'Sample mean Xbar =' and 'Sample SD s ='. To input the given information, in cell C4, type the corresponding value **10**. In cell D4, type the label 'Data' and type the given 10 family food expenditures in cells D5 to D14. To obtain the sample mean $\overline{X}$ and the sample standard deviation s from the data set, in cells C5 and C6, type the formulas **=AVERAGE(D5:D14)** and **=STDEV(D5:D14).**

Task 1 The null and the alternative hypotheses are

$$H_0: \quad \mu = 7$$
$$H_A: \quad \mu > 7$$

As the alternative hypothesis contains the 'greater than' sign, the rejection region of the test is in the right hand tail and this is a one-tail test.

In cells B8 and B9, type the labels 'Ho: Mu =' and 'HA: Mu > ' and, in cells C8 and C9, type the values **7** and **7**, respectively.

Task 2 As the population variable X is normally distributed, $\overline{X}$ is distributed exactly normal. When the population standard deviation σ is unknown and estimate it by the sample standard deviation s, to test a hypothesis about the population mean μ, we use the test statistic (see Case 2, Section 10.2)

$$t = \frac{\overline{X} - \mu}{s/\sqrt{n}}$$

based on a t-distribution with (n-1) = 9 degrees of freedom under the assumption that the population is normal.

In cells B11 and C11, type the labels 'Test Statistic:' and 't = (Xbar – mu)/(s/sqrt(n))'.

Task 3 In cell B13, type the label 'Level of significance =' and, in cell C13, type the given value **0.10**.

Task 4 ***Calculating the critical value(s)***

As this is a *(right) one-tail test* (see diagram below), for a test with 10 percent level of significance,

$$\text{Critical value } t_{critical} = t_{\alpha,n-1} = t_{.10,9}$$

This value is based on a t-distribution and can be obtained using the EXCEL function TINV. In cell B15, type 'Critical value t(.10,9) ='. To obtain the critical value, in cell C15, type the formula **=TINV(.2,9)** or **=TINV(2*C13,C4-1)**. This will give a value of 1.383029 in cell C15. [Note that TINV calculates critical values for two-tailed probabilities only. Therefore, we use probability .2 to obtain the t-value corresponding to a one-tail probability of .1. You should convince yourself by comparing the t-value in cell C15 (1.383029) with that from the t-table for $t_{.10,9}$ (= 1.383).]

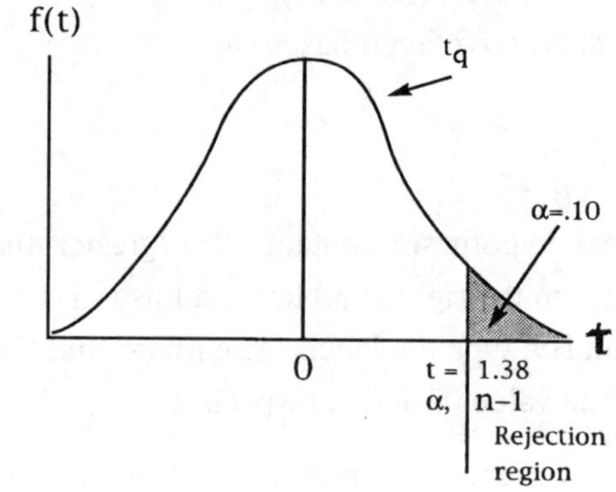

Alternatively, you can click on **Paste Function** and then select *Function Category* **Statistical**. From the *Function Name* select **TINV** and click **OK** on the dialogue box. The **TINV** dialogue box will now appear on the screen (see Figure 10.8). This will ask for two inputs, namely, *Probability* (for this task, type **.2** or **2*C13**) and *Deg_freedom* (for this task, type the degrees of freedom **9 or C4-1**). The critical value ($t_{critical}$) of 1.383029 will appear in cell C15.

Figure 10.8 TINV Function dialogue box

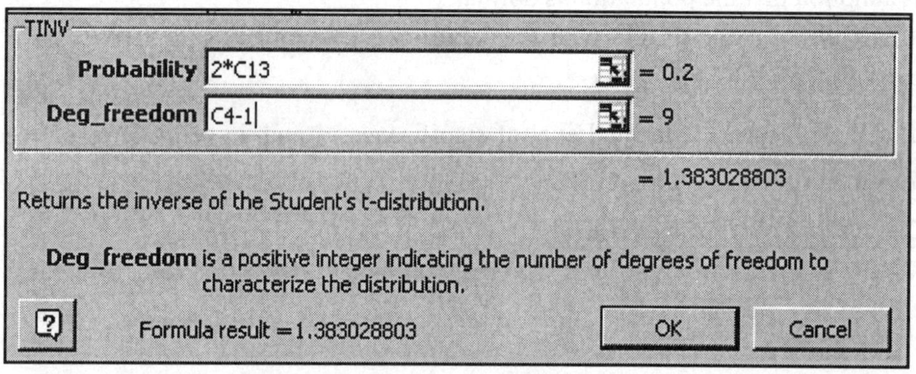

The decision rule

Therefore the decision rule for the test is

Method A: **Reject H_0 if $t_{calculated} > t_{critical} = 1.383029$**

Method B: **Reject Ho if *p*-value < α = .10**

In cells B16 to B18, type the labels 'Decision Rule: ', 'Method A: Reject Ho if tcal >' and 'Method B: Reject Ho if p-value <'. In cells C17 and C18, type the formulas **=C15** and **=C13**.

Task 5 ***Calculating the data-based value of the test statistic ($t_{calculated}$)***

In cells B20 and B21, type the labels 'SE of sample mean =', 'Test statistic t-calculated ='. To obtain the standard error of the sample mean, in cell C20, type the formula **=C6/SQRT(C4)**.

To calculate the data-based value of the test statistic, in cell C21, type the formula **=(C5-C8)/C20**. Now this will give the value (=1.81249) for the test statistic calculated above.

Calculating the p-value

For a right-tail t-test, the *p*-value = $P(t > t_{calculated})$, where $t_{calculated}$ is the standardised value of the test statistic.

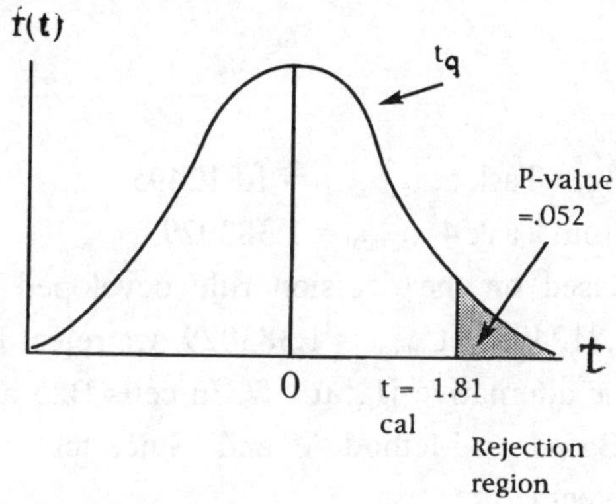

In cells B22 and B23, type the labels 'Tail probability' and 'p-value=' and click the mouse on C22. In this case, we use *Function Name* **TDIST** under the *Function Category* **Statistical** of **Paste Function** to give tail area. The TDIST function returns the tail area corresponding to a particular t value. When you select **TDIST** and click **OK**, the dialogue box for **TDIST** appears on the screen (see Figure 10.9) and will ask for 3 inputs, namely, *X* (the value of the test statistic t under H_0, which must be given in its absolute form as this argument should always be positive and TDIST always returns the right tail area when you assign number of tails = 1), *Deg_freedom*, and the number of *Tails*. In this example, the three inputs are **ABS(C21)**; **9** or **(C4-1)**; and **1**. Click **OK**. This will give you a tail probability value of .051663 in cell C22. In cell C23, type the formula **=C22**, to obtain the *p*-value.

Alternatively, type the formula = **TDIST(ABS(C21),C4-1,1)** in cell C23.

[Note that for a two-tail test under a t-distribution, the *p*-value = TDIST(ABS($t_{calculated}$),degrees of freedom,2) or 2*TDIST(ABS($t_{calculated}$), degrees of freedom,1).]

Figure 10.9 TDIST Function dialogue box

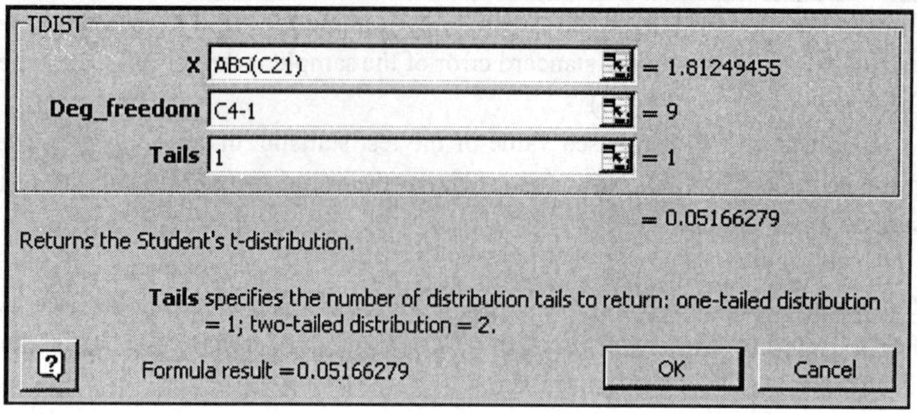

Task 6 ***Conclusion***

(a) Method A From Task 5, $t_{calculated}$ = 1.812495

From Task 4, $t_{critical}$ = 1.383029.

Based on the decision rule developed in Task 4, $t_{calculated}$ = 1.812495 > $t_{critical}$ = 1.383029, we reject H_O: $\mu = 7$ in favour of the alternative H_A: $\mu > 7$. In cells B25 and C25, type the text, 'Based on Method A' and 'Since tcal = 1.81 > tcrit = 1.383, reject Ho'.

(b) Method B From Task 5, *p*-value = .051663

Based on the decision rule given in Task 4, as the level of significance $\alpha = .10 > p$-value, we reject H_O: $\mu = 7$ in favour of the alternative H_A: $\mu > 7$. In cells B26 and C26, type the text, 'Based on Method B' and 'Since p-value = .051663 < alpha = .10, reject Ho'.

Task 7 Save your workbook as **A:EX104.XLS** on your disk and print the file. The output from Tasks 1-6 is given in Figure 10.10.

Figure 10.10 Hypothesis testing about μ when σ is unknown

	A	B	C	D
1		Hypothesis Testing about Population Mean Mu		
2		Population Standard Deviation Sigma is Unknown		
3				
4	Data Summary:	Sample size n =	10	Data
5		Sample mean Xbar =	8.9	4
6		Sample SD s =	3.314949304	8
7				12
8	Step 1	Ho: Mu =	7	11
9		HA: Mu >	7	14
10				6
11	Step 2	Test statistic	t=(Xbar - mu)/(s/sqrt(n))	12
12				8
13	Step 3	Level of significance =	0.1	9
14				5
15	Step 4	Critical value t(.10,9) =	1.383028803	
16		Decision rule:		
17		Method A: Reject Ho if tcal >	1.383028803	
18		Method B: Reject Ho if p-value <	0.1	
19				
20	Step 5	SE of sample mean =	1.048279013	
21		Test statistic t-calculated =	1.812494552	
22		Tail probability =	0.05166279	
23		p-value =	0.05166279	
24				
25	Step 6	Based on Method A	Since tcal =1.81 > tcrit = 1.38, reject Ho	
26		Based on Method B	Since p-val = .051 < alpha = .10, reject Ho	

10.7 Example 10.5: Testing hypotheses about the population mean μ when σ is unknown and n > 30

(Ref: Based on Exercise 10.20, p. 120, Australian Business Statistics Study Guide, Selvanathan et al., 1995)

An oil refiner has developed a new type of oil that is supposed not only to reduce engine wear but also increase fuel economy. A random sample of 50 cars that have a mean performance of 8 kilometers per litre is selected. When the new oil is put into the cars, each is driven until the tank is dry. The number of kilometers per litre is recorded. The mean and standard deviation of the 50 observations are $\overline{X} = 8.8$ and s=2.5. At the 5% significance level, can we conclude that the new type of oil is effective in increasing fuel economy? Assume that the population is normally distributed. Perform the following tasks as intermediate steps to reach your conclusion.

Tasks 1-6 Perform the same Tasks 1-6 as in the Example 10.1 question. Use EXCEL to perform Tasks 4 and 5. Include in your EXCEL file information regarding the other tasks.

Task 7 Save your workbook as 'A:EX105.XLS' on your disk and print the file.

Solution to Example 10.5

One major difference between this example and the last example is that, in the last example, the sample size was only 10, Here, it is larger than 30. Therefore we can use the standard normal approximation to the t-distribution.

Let X be the number of kilometers per litre recorded for a car, μ be the population mean (in kilometers per litre) and σ be the population standard deviation. In cell A1, type the title 'Hypothesis Testing about Population Mean Mu' and center it across cells Al to D1. Type the second title 'Population Standard Deviation Sigma is Unknown and n > 30' in cell A2 and center it across cells A2 to D2.

Data summary (given)

Sample size n = 50; sample mean $\overline{X}$ = 8.8; population standard deviation σ (unknown) and estimated by the sample standard deviation s = 2.5; significance level α = .05; population variable X, the number of kilometers per litre recorded for a car, is normally distributed. In cells B4 to B6, type the labels 'Sample size n =', 'Sample mean Xbar =' and 'Sample SD s ='. To input the given information, in cells C4 to C6, type the corresponding values **50, 8.8** and **2.5.**

Task 1 The null and the alternative hypotheses are

$$H_0: \quad \mu = 8$$
$$H_A: \quad \mu > 8$$

As the alternative hypothesis contains the 'greater than' sign, the rejection region of the test is in the right hand tail and this is a one-tail test.

In cells B8 and B9, type the labels 'Ho: Mu ='and 'HA: Mu > ' and, in cells C8 and C9, type the values **8** and **8**.

Task 2 As the population variable X is normally distributed, $\overline{X}$ is distributed exactly normal. Therefore, when the population standard deviation σ is *unknown* and n > 30, to test a hypothesis about the population mean μ, we use the test statistic (see Case 3, Section 10.2)

$$Z = \frac{\overline{X} - \mu}{s/\sqrt{n}}$$

which has an approximate z-distribution with degrees of freedom. [As the degrees of freedom n-1 =49 >30, we have approximated the t-distribution with a standard normal distribution.]

In cells B11 and C11, type the labels 'Test Statistic:' and 'Z = (Xbar – mu)/(s/sqrt(n))'.

Task 3 In cell B13, type the label 'Level of significance =' and, in cell C13, type the given value **0.05**.

Task 4 ***Calculating the critical value(s)***

As this is a (right) one-tail test, for a test with 10 percent level of significance,

$$\text{Critical value} = z_{\alpha} = z_{.05}$$

This value is based on a standard normal distribution and can be obtained using the EXCEL function NORMSINV. In cell B15, type 'Critical value t(.05,49)=Z(.05)'. As before in Task 4 of Example 10.1, to obtain the critical value, in cell C15, type the formula **=NORMSINV(.95)** or **=NORMSINV(1-C13)**. The critical value will be 1.644853.

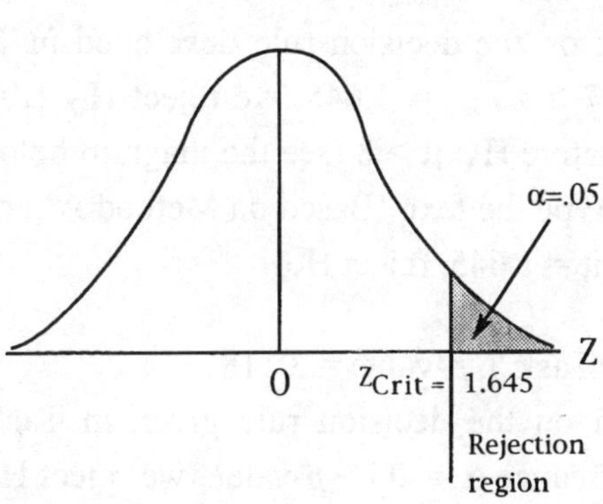

The decision rule

Therefore, the decision rule for the test is

Method A: **Reject H_0 if $z_{calculated} > z_{critical} = 1.644853$**

Method B: **Reject Ho if *p*-value < $\alpha = .05$**

In cells B16 to B18, type the labels 'Decision Rule: ', 'Method A: Reject Ho if Zcal >' and 'Method B: Reject Ho if p-value <'. In cells C17 and C18, type the formulas **=C15** and **=C13**.

Task 5 ***Calculating the data-based value of the test statistic ($z_{calculated}$)***

In cells B20 and B21, type the labels 'SE of sample mean =' and 'Test statistic Z-calculated ='. To obtain the standard error of sample mean, in cell C20, type the formula **=C6/SQRT(C4)**.

To calculate the data-based value of the test statistic, in cell C21, type the formula **=(C5-C8)/C20**. This will give the value (=2.2627417) for the test statistic.

Calculating the p-value

For a right-tail z-test, the *p*-value = $P(Z > z_{calculated})$, where $z_{calculated}$ is the standardised value of the test statistic.

To calculate this probability, first type the label 'Cumulative probability =' in cell B22 and select cell C22 to store the cumulative probability corresponding to $z_{calculated}$ = 2.2627 (the sample value of the test statistic calculated above). Thus, the *p*-value for a (right) one-tail test will be one minus this cumulative probability. Now we use the function **NORMSDIST** or **NORMDIST** to obtain the cumulative probability corresponding to the Z value = 2.2627 and then get the *p*-value as (1-cumulative probability) since H_A: μ > 8, we use the right tail. That is, in cell C22, type the formula **=NORMSDIST(2.2627)** or **=NORMSDIST(C21)**. To calculate the *p*-value, in cell B23, type the label 'p-value=' and, in cell C23, type the formula **=1-C22**.

Task 6 ***Conclusion***

(a) Method A From Task 5, $z_{calculated}$ = 2.2627.
From Task 4, $z_{critical}$ = 1.645.
Based on the decision rule developed in Task 4, as $z_{calculated}$ = 2.2627 > $z_{critical}$ = 1.645, we reject H_O: μ = 8 in favour of the alternative H_A: μ > 8 (see the diagram below). In cells B25 and C25, type the text, 'Based on Method A' and 'Since Zcal = 2.26 > Zcrit = 1.645, reject Ho'.

(b) Method B From Task 5, *p*-value = .0118.
Based on the decision rule given in Task 4, as the level of significance α = .05 > *p*-value, we reject H_O: μ = 8 in favour of the alternative H_A: μ > 8 (see the diagram below). In cells B26 and C26, type the text, 'Based on Method B' and 'Since p-value = .0118 < alpha = .05, reject Ho'.

Method A

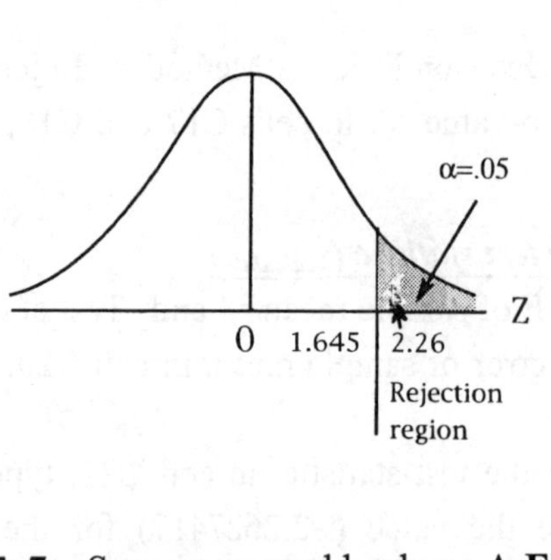

Method B

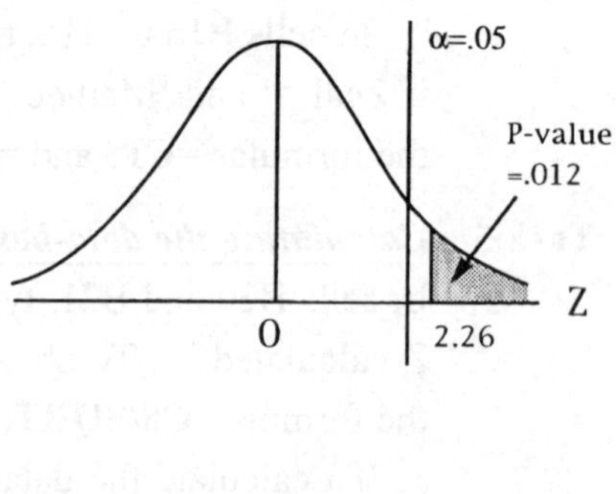

Task 7 Save your workbook as **A:EX105.XLS** on your disk and print the file. The output from Tasks 1-6 is given in Figure 10.11.

Figure 10.11 Hypothesis testing about μ when σ is unknown and n > 30

	A	B	C
1		**Hypothesis Testing about Population Mean Mu**	
2		**Population Standard Deviation Sigma is Unknown and n >30**	
3			
4	Data Summar	Sample size n =	50
5		Sample mean Xbar =	8.8
6		Sample SD s =	2.5
7			
8	Step 1	Ho: Mu =	8
9		HA: Mu >	8
10			
11	Step 2	Test statistic	Z=(Xbar - mu)/(s/sqrt(n))
12			
13	Step 3	Level of significance =	0.05
14			
15	Step 4	Critical value t(.05,49) = Z(.05)	1.644853
16		Decision rule:	
17		Method A: Reject Ho if Zcal >	1.644853
18		Method B: Reject Ho if p-valu	0.05
19			
20	Step 5	SE of sample mean =	0.353553391
21		Test statistic Z-calculated =	2.2627417
22		Cumulative probability =	0.988174228
23		p-value =	0.011825772
24			
25	Step 6	Based on Method A	Since Zcal =2.26 > Zcrit = 1.645, reject Ho
26		Based on Method B	Since p-val = .011 < alpha = .05, reject Ho

10.8 Hypothesis testing involving a single population: Qualitative data type

When the data type is qualitative, the problem objective is to test hypotheses about the population proportion p. To test hypotheses involving p, we follow the same six-step procedure described in Section 10.1.

In this case, the null and the alternative hypotheses will take the following forms:

$$H_o:\quad p = p_o$$
$$H_A:\quad p \neq p_o \quad \text{or} \quad p < p_o \quad \text{or} \quad p > p_o$$

where p_o is the value of the population proportion p under the null hypothesis.

The test statistic is

$$Z = \frac{\hat{p} - p_o}{\sqrt{\dfrac{p_o q_o}{n}}} \sim z(0,1)$$

provided $n\hat{p} \geq 5$ and $n\hat{q} \geq 5$, n is the sample size and $q_o = 1 - p_o$.

10.9 Example 10.6: Testing hypotheses about the population proportion (p) – one-tail test

(Ref: Based on Exercise 10.43, p. 294, Australian Business Statistics, Selvanathan et al., 1994)

A tyre manufacturer claims that more than 90% of her tyres will last at least 80,000km. In a random sample of 200 tyres, 10 wore out before reaching 80,000km. Do the data support the manufacturer's claim, with a level of significance $\alpha = .01$? Perform the following tasks as intermediate steps to reach your conclusion.

Tasks 1-6 Perform the same Tasks 1-6 as in the Example 10.1 question. Use EXCEL to perform Tasks 4 and 5. Include in your EXCEL file information regarding the other tasks.

Task 7 Save your workbook as 'A:EX106.XLS' on your disk and print the file.

Solution to Example 10.6

Let X be the number of tyres lasting at least 80,000km and p be the corresponding proportion of tyres lasting at least 80,000km in the population. Then$p$ = X/N, where N is the population size. Let $\hat{X}$ be the number of tyres lasting at least 80,000km and $\hat{p}$ be the corresponding proportion of tyres lasting at least 80,000km in the sample. Then $\hat{p} = \hat{X}/n$, where n is the sample size. In cell A1, type the title 'Hypothesis Testing for Population Proportion p', and center it across cells A1 to F1.

Data summary (given)

Sample size $n = 200$; $\hat{X} = 200\text{-}10 = 190$; significance level $\alpha = .01$. In cells B4 to B6, type the labels 'Sample size n =', 'Sample Xhat =' and 'Sample proportion phat='. To input the given information, in cells C4 and C5, type the corresponding values **200** and **190**. To obtain the value of $\hat{p}$ which is $\hat{X}/n$, in cell C6, type the formula **=C5/C4**.

Task 1 We want to test the claim that more than 90% of her tyres will last at least 80,000km. Therefore, the null and the alternative hypotheses are

$$H_0: \quad p = .90$$
$$H_A: \quad p > .90$$

As the alternative hypothesis contains the 'greater than' sign, the rejection region of the test is in the right hand tail and this is a one-tail test.

In cells B8 and B9, type the labels 'Ho: p =' and 'HA: p > ' and, in cells C8 and C9, type the values, **.9** and **.9**.

Task 2 When $n\hat{p} \geq 5$ and $n\hat{q} \geq 5$, to test a hypothesis about the population proportion, we use the test statistic (see Section 10.6)

$$Z = \frac{\hat{p} - p}{\sqrt{\frac{pq}{n}}}$$

based on the standard normal distribution.

In cells B11 and C11, type the labels 'Test Statistic:' and 'Z = (phat – p)/sqrt(pq/n)'.

Task 3 In cell B13, type the label 'Level of significance =' and, in cell C13, type the given value **0.01**.

Task 4 ***Calculating the critical value(s)***

As this is a (right) one-tail test (see diagram below), for a test with 1 percent level of significance,

$$\text{Critical value } z_{critical} = z_\alpha = z_{.01}.$$

This value is based on a standard normal distribution and can be obtained using the EXCEL function NORMSINV. In cell B15, type the label 'Critical value z(.01) ='. Using the same procedure as in Task 4 of Example 10.1, to obtain the critical value, in cell C15, type the formula **=NORMSINV(.99)** or **=NORMSINV(1-C13)**. The critical value will be 2.326342.

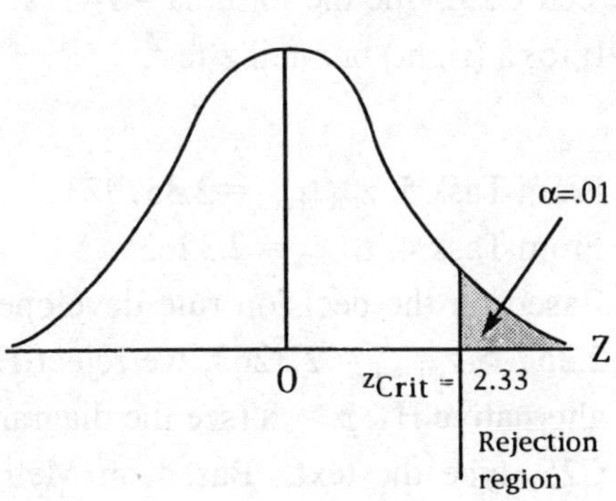

The decision rule

Therefore the decision rule for the test is (see diagram below)

Method A: **Reject H_0 if $z_{calculated} > z_{critical} = 2.326342$**
Method B: **Reject Ho if *p*-value $< \alpha = .01$**

In cells B16 to B18, type the labels 'Decision Rule: ', 'Method A: Reject Ho if Zcal >' and 'Method B: Reject Ho if p-value <'. In cells C17 and C18, type the formulas **=C15** and **=C13**.

Task 5 ***Calculating the data-based value of the test statistic ($z_{calculated}$)***

In cells B20 and B21, type the labels 'SE of phat=' and 'Test statistic Z-calculated ='. To obtain the standard error of phat in cell C20, type the formula **=SQRT(C8*(1-C8)/(C4))**.

Select cell C21 to store the sample value of the test statistic.

Method 1 To calculate the value of the test statistic, in cell C21, type the formula **=(C6-C8)/C20**.

Method 2 You can also obtain this z-value, by typing the formula **=STANDARDIZE(C6,C8,C20)** in cell C21.

This will give the value (= 2.357023) of the test statistic in cell C21.

Calculating the p-value

For a right-tail z-test, the *p*-value = $P(Z > z_{calculated})$, where $z_{calculated}$ is the standardised value of the test statistic.

To calculate this probability, we use the same procedure as in calculating the *p*-value in Task 4 of Example 10.1. First type the label 'Cumulative probability =' in cell B22 and then, in cell C22, type the formula **=NORMSDIST(2.357023)** or **NORMSDIST(C21)** to obtain cumulative probability corresponding to $z_{calculated}$ = 2.357023 (the sample value of the test statistic calculated above). Now, the *p*-value is (1-cumulative probability), since H_A: $p > .9$, we use the right tail. Thus, to obtain the *p*-value, in cell B23, type 'p-value =' and, in cell C23, type the formula = **1-C22**. This will give the required *p*-value, .009211, for a (right) one-tail z-test.

Task 6 ***Conclusion***

(a) Method A From Task 5, $z_{calculated}$ = 2.357023

From Task 4, $z_{critical}$ = 2.3263

Based on the decision rule developed in Task 4, as $z_{calculated}$ = 2.357 > $z_{critical}$ = 2.3263, we reject H_O: p = .9 in favour of the alternative H_A: $p > .8$ (see the diagram below). In cells B25 and C25, type the text, 'Based on Method A' and 'Since Zcal = 2.357 > Zcrit = 2.326, reject Ho'.

(b) Method B From Task 5, *p*-value = .009211

Based on the decision rule given in Task 4, as the level of significance α = .01 > *p*-value, we reject H_O: p = .9 in favour of the alternative H_A: $p > .9$ (see the diagram below). In cells B26 and C26, type the text, 'Based on Method B' and 'Since p-value = .009 < alpha = .01, reject Ho'.

Task 7 Save your workbook as **A:EX106.XLS** on your disk and print the file. The output from Tasks 1-6 is given in Figure 10.12.

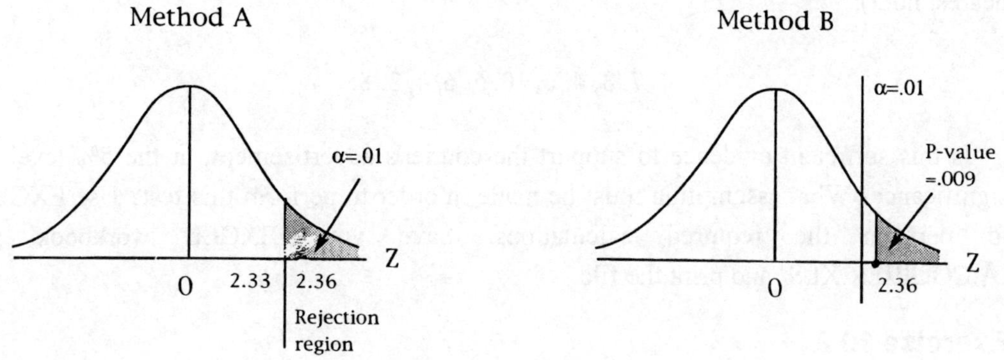

Figure 10.12 Hypothesis testing about p

	A	B	C	D	E	F
1		Hypothesis Testing for Population Proportion p				
2						
3						
4	Data Summary:	Sample size n =	200			
5		Sample Xhat =	190			
6		Sample proportion phat =	0.95			
7						
8	Step 1	Ho: p =	0.9			
9		HA: p >	0.9			
10						
11	Step 2	Test statistic	Z=(phat - p)/sqrt(pq/n)			
12						
13	Step 3	Level of significance =	0.01			
14						
15	Step 4	Critical value z(.01) =	2.326342			
16		Decision rule:				
17		Method A: Reject Ho if Zcal >	2.326342			
18		Method B: Reject Ho if p-value <	0.01			
19						
20	Step 5	SE of phat =	0.021213			
21		Test statistic Z-calculated =	2.357023			
22		Cumulative probability =	0.990789			
23		p-value =	0.009211			
24						
25	Step 6	Based on Method A	Since Zcal = 2.357 > Zcrit = 2.326, Reject Ho			
26		Based on Method B	Since p-val = .009 < alpha = .01, Reject Ho			

10.10 Exercises to Chapter 10

(The following exercises are based on exercises from Chapter 10 of Selvanathan et al., Australian Business Statistics, 1994)

Exercise 10.1

A courier service in Brisbane advertises that its average delivery time is less than 6 hours for local deliveries. A random sample of the amount of time this courier takes to deliver packages to an address across town produced the following delivery times (rounded to the nearest hour):

7, 3, 4, 6, 10, 5, 6, 4, 3, 8.

Is this sufficient evidence to support the courier's advertisement, at the 5% level of significance? What assumption must be made in order to perform this test? Use EXCEL to perform the required calculations. Save your EXCEL workbook as 'A:COURIER.XLS' and print the file.

Exercise 10.2

A machine that produces ball bearings is set so that the average diameter is 1cm. In a sample of 100 ball bearings, it was found that $\overline{X}$ = 1.02cm Assuming that the standard deviation is .1cm, can we conclude (with α = .05) that the mean diameter is not 1cm? Use EXCEL to perform the required calculations. Save your EXCEL workbook as 'A:BALL.XLS' and print the file.

Exercise 10.3

A university in Queensland claims that the average TE score of applicants to its Business program has increased during the past three years. Three years ago the mean and the standard deviation of TE scores of the university's Business program applicants were 920 and 20 respectively. In a sample of 36 of this year's applicants for the Business program, the mean TE was 925. At the 5% level of significance, can we conclude that the university's claim is true? (Assume that the standard deviation is unchanged). Use EXCEL to perform the required calculations. Save your EXCEL workbook as 'A:TESCORE.XLS' and print the file.

Exercise 10.4

An automotive expert claims that the large number of self-serve petrol stations has resulted in poor automobile maintenance, and that the average tyre pressure is at least 4 pounds per square inch (psi) below its manufacturer's specification. As a quick test, ten tyres are examined, and the number of pounds per square inch each tyre is below specification is recorded. The resulting data (in psi) are as follows:

7, 9, 2, 0, 5, 6, 3, 5, 8, 9.

Is there sufficient evidence, with α = .05, to support the expert's claim? Use EXCEL to perform the required calculations. Save your EXCEL workbook as 'A:AUTO.XLS' and print the file.

Exercise 10.5

A fast food franchiser in Strathfield is considering building a restaurant at a certain location. Based on financial analysis, a site is acceptable only if the number of pedestrians passing the location averages at least 100 per hour. A random sample of 50 hours produced $\overline{X}$ = 110 and s = 12 pedestrians per hour. Do these data provide sufficient evidence to establish that the site is acceptable? (Use $\alpha = .05$). Use EXCEL to perform the required calculations. Save your EXCEL workbook as 'A:FRANCH.XLS' and print the file.

Exercise 10.6

In a television commercial, the manufacturer of a toothpaste claims that 4 out 5 dentists recommend its product. In order to test the claim, a consumer-protection group randomly samples 400 dentists and finds that 310 recommended that toothpaste. At the 5% level of significance, test to determine if sufficient evidence exists to refute the manufacturer's claim. (Interpret the claim to say that at least 4 out of 5 dentists recommend the product.) Use EXCEL to perform the required calculations. Save your EXCEL workbook as 'A:TOOTH.XLS' and print the file.

Chapter 11

Hypothesis Testing: Comparing Two Populations

In the previous chapter we looked at hypothesis testing about population parameters involving a single population. In this chapter, we discuss how to perform hypothesis testing when the problem objective is to compare two populations. When the data type is quantitative, we will test hypotheses about the difference between the two population means, and when the data type is qualitative, we will test hypotheses about the difference between the two population proportions. For this purpose, we use the same *six-step procedure* and methods applied in the previous chapter.

Aim: At the completion of the examples in this Chapter, you should be able to use EXCEL to perform hypothesis testing involving two populations about the

(i) population means μ_1 and μ_2 when the population variances σ_1^2 and σ_2^2 are known;

(ii) population means μ_1 and μ_2 when the population variances σ_1^2 and σ_2^2 are unknown;

(iii) population means μ_1 and μ_2 when the population variances σ_1^2 and σ_2^2 are unknown but are equal;

(iv) population means μ_1 and μ_2 when the population variances σ_1^2 and σ_2^2 are unknown and sample sizes are greater than 30; and

(v) population proportions p_1 and p_2.

11.1 Hypothesis testing involving two populations: Quantitative data type

When making inferences about two populations where the data type is quantitative, the problem objective is to test hypotheses about comparing the population means. The parameter of interest is the difference between the population means (μ_1 - μ_2). To test hypotheses involving (μ_1 - μ_2), we follow the same six-step procedure described in Section 10.1.

In this case, the null and the alternative would take the following forms:

H_0: $\mu_1 - \mu_2 = D$

H_A: $\mu_1 - \mu_2 \neq D$ (two tail) *or*
$\mu_1 - \mu_2 < D$ (left one tail) *or*
$\mu_1 - \mu_2 > D$ (right one tail)

where D is any real value. Usually, we test for the equality of the two population means. In this case, D = 0. Assuming that the two samples are independent, the test statistic takes one of the following forms.

Case 1 ***When both population variances σ_1^2 and σ_2^2 are known***

$$z = \frac{\left(\overline{X}_1 - \overline{X}_2\right) - \left(\mu_1 - \mu_2\right)}{\sqrt{\frac{\sigma_1^2}{n_1} + \frac{\sigma_2^2}{n_2}}} \sim \text{Normal}(0,1)$$

provided either both populations are normally distributed or if the distribution of the populations are unknown or non-normal and the sample sizes n_1, $n_2 \geq 30$ (using the Central Limit theorem).

Case 2 ***When the population variances σ_1^2 and σ_2^2 are unknown***

$$t = \frac{\left(\overline{X}_1 - \overline{X}_2\right) - \left(\mu_1 - \mu_2\right)}{\sqrt{\frac{s_1^2}{n_1} + \frac{s_2^2}{n_2}}} \sim t_\nu$$

provided both populations are normally distributed, where the degrees of freedom

$$\nu = \frac{\left[\frac{s_1^2}{n_1} + \frac{s_2^2}{n_2}\right]^2}{\frac{\left(s_1^2 / n_1\right)^2}{n_1 - 1} + \frac{\left(s_2^2 / n_2\right)^2}{n_2 - 1}}$$

Case 3 ***When the population variances σ_1^2 and σ_2^2 are unknown but are equal***

$$t = \frac{\left(\overline{X}_1 - \overline{X}_2\right) - \left(\mu_1 - \mu_2\right)}{\sqrt{\left(\frac{s_P^2}{n_1} + \frac{s_P^2}{n_2}\right)}} \sim t_{n_1+n_2-2}$$

provided both populations are normally distributed, where the pooled variance

$$s_P^2 = \frac{(n_1 - 1)s_1^2 + (n_2 - 1)s_2^2}{n_1 + n_2 - 2}$$

Case 4 ***When the population variances σ_1^2 and σ_2^2 are unknown and $n_1, n_2 > 30$***

$$z = \frac{\left(\overline{X}_1 - \overline{X}_2\right) - \left(\mu_1 - \mu_2\right)}{\sqrt{\frac{s_1^2}{n_1} + \frac{s_2^2}{n_2}}} \sim t_v \approx \text{Normal}(0,1)$$

where ν is the degrees of freedom, provided both populations are normally distributed and the sample sizes n_1, $n_2 > 30$.

Note that the difference between Case 2 and Case 4 is that, in Case 4, the sample size is larger than 30 and therefore the t-distribution can be approximated by the z-distribution.

In EXCEL, when testing hypotheses where the raw data are given, there are two ways of performing the necessary steps (i.e., obtaining the critical value(s), the data-based value of the test statistic, the p-value etc.). In Method 1, we use the formulas to calculate the necessary values. In Method 2, we use the **Data Analysis** tool to perform the calculations automatically. When testing hypotheses where only the summary data (i.e., sample mean, sample standard deviation) are given, Method 2 is not suitable. In this situation, we can use Method 1 only. In the following Examples 11.1-11.3, as the raw data are given, we illustrate both methods.

11.2 Example 11.1: Testing hypotheses about the difference between two population means D = (μ_1 - μ_2) when D = 0 and σ_1^2 and σ_2^2 are known

(Ref: Based on Example 11.1, p. 128, Australian Business Statistics, Study Guide, Selvanathan et al., 1995)

A study of the scholastic aptitude test (SAT) revealed that in a random sample of 6 males, the SAT scores were 290, 418, 507, 469, 510 and 349.4, and in a random sample of 8 females, the SAT scores were 260, 330, 445, 520, 490, 535, 420 and 452. If the populations are normally distributed with standard deviations 95 and 85 for male and female scores, respectively, can we conclude at the 1% significance level that male scores are lower than the female scores? Perform the following tasks as intermediate steps to reach your conclusion. Use EXCEL to perform Tasks 4 and 5. Include in your EXCEL file information regarding the other tasks.

Task 1 Determine the null and the alternative hypotheses.
Task 2 Determine the test statistic and its sampling distribution.
Task 3 Specify the level of significance.
Task 4 Obtain the critical value(s) and develop the decision rule.
Task 5 Obtain the sample value of the test statistic and the corresponding *p*-value.
Task 6 Derive your conclusion for the test in two ways:

(a) using the sample value of the test statistic obtained in Task 5 and the decision rule developed in Task 4.

(b) using the *p*-value obtained in Task 5 and the level of significance given in Task 3.

Task 7 Save your workbook as 'A:EX111.XLS' and print the file.

Solution to Example 11.1

Let X_1 and X_2 be the SAT scores of male and female populations, μ_1 and μ_2 be the population mean scores, and σ_1 and σ_2 be the population standard deviations, respectively. In cell A1, type the title 'Hypothesis Testing about Population Mean Difference D = (Mu1 – Mu2) when D = 0' and center it across cells A1 to D1. Type the second title 'Population Standard Deviations Sigma1 and Sigma2 are Known' in cell A2 and center it across cells A2 to D2. In cells E3 and F3, type the headings, 'Male SAT' and 'Female SAT', respectively. In cells E4 to E9, type the 6 male scores and, in cells F4 to F11, type the 8 female scores.

Note: In this example, as we have access to raw data, we can use either Method 1 or Method 2 (described below) to perform Tasks 4 and 5.

Data summary (given)

Sample sizes $n_1 = 6$, $n_2 = 8$; population standard deviations are known and $\sigma_1 = 95$ and $\sigma_2 = 85$; and significance level $\alpha = .01$. In cells B4 to B6, type the labels 'Sample size n =', 'Sample mean Xbar =' and 'Population SD sigma ='. To input the given information, in cells C4 and C6, type the corresponding Sample 1 values **6** and **95** and, in cells D4 and D6, type the corresponding Sample 2 values **8** and **85**. In cells C5 and D5, type the formulas **=AVERAGE(E4:E9)** and **=AVERAGE(F4:F11)**.

Method 1

Task 1 To test whether male scores are lower than female scores, we set up the null and alternative hypotheses as follows:

H_0: $\mu_1 - \mu_2 = 0$
H_A: $\mu_1 - \mu_2 < 0$

As the alternative hypothesis contains the 'less than' sign, the rejection region of the test is in the left hand tail and this is a one tail test.

In cells B8 and B9, type the labels 'Ho: (Mu1 – Mu2) =', 'HA: (Mu1 – Mu2) < ' and, in cells C8 and C9, type the values, **0** and **0**.

Task 2 As the population standard deviations σ_1 and σ_2 are <u>*known*</u> and the populations are normally distributed, to test a hypothesis about the difference between two population means ($\mu_1 - \mu_2$), we use the test statistic (see <u>Case 1</u>, Section 11.1)

$$z = \frac{(\bar{X}_1 - \bar{X}_2) - (\mu_1 - \mu_2)}{\sqrt{\frac{\sigma_1^2}{n_1} + \frac{\sigma_2^2}{n_2}}} \sim \text{Normal}(0,1)$$

That is, we use the standard normal distribution.

In cells B11 and C11, type the labels 'Test Statistic:' and 'Z = [(Xbar1 – Xbar2) – (Mu1 – Mu2)]/SQRT[(sigma1^2/n1)+(sigma2^2/n2)]]'.

Task 3 In cell B13, type the label 'Level of significance =' and, in cell C13, type the given value **0.01**.

Task 4 <u>***Calculating the critical value***</u>

As this is a (left) one tail test (see diagram below), for a test with 1 percent level of significance,

$$\text{Critical value } z_{critical} = -z_\alpha = -z_{.01}.$$

This value is based on a standard normal distribution and can be obtained using the EXCEL function NORMSINV. In cell B15, type the label 'Critical value z(.01) ='. The critical region is on the left tail of the z-distribution with area .01. To obtain this value, in cell C15, type **=NORMSINV(.01)** or **=NORMSINV(C13).** Now the $z_{critical}$ value of –2.3263 will appear in cell C15.

<u>***The decision rule***</u>

Therefore the decision rule for the test is

Method A: **Reject H_0 if $z_{calculated} < z_{critical} = -2.326$**
Method B: **Reject H_0 if *p*-value $< \alpha = .01$**

In cells B16 to B18, type the labels 'Decision Rule: ', 'Method A: Reject Ho if Zcal <' and 'Method B: Reject Ho if p-value <'. In cells C17 and C18, type the cell references **=C15** and **=C13,** respectively.

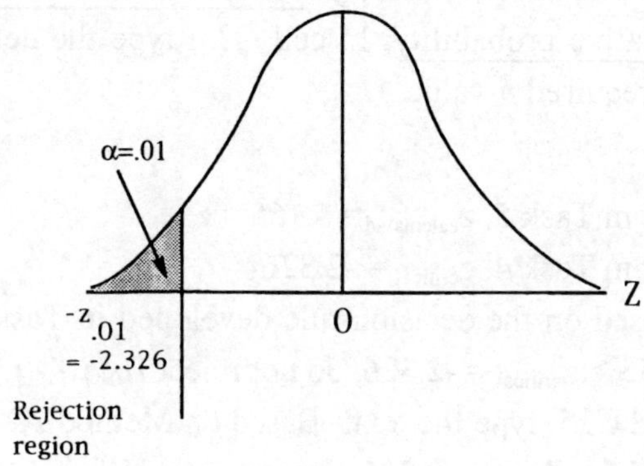

Task 5 ***Calculating the data-based value of the test statistic ($z_{calculated}$)***

To calculate the data-based value of the test statistic defined in Task 2, first we calculate the denominator which is the standard error of ($\bar{X}_1$ - $\bar{X}_2$) and then calculate the data-based value of the test statistic ($z_{calculated}$).

In cells B20 to B21, type the labels 'SE of sample mean =' and 'Test statistic Z-calculated ='. To obtain the standard error, in cell C20, type the formula **=SQRT((C6^2/C4)+(D6^2/D4))**. Now, to calculate $z_{calculated}$, in cell C21, type the formula **=(C5-D5-C8)/C20** or type **=STANDARDIZE(C5-D5,C8,C20)**. This will give the $z_{calculated}$ value (= -.155) of the test statistic in cell C21.

Calculating the p-value

For a (left) one tail z-test, the *p*-value = P(Z < $z_{calculated}$), where $z_{calculated}$ is the standardised value of the test statistic calculated above.

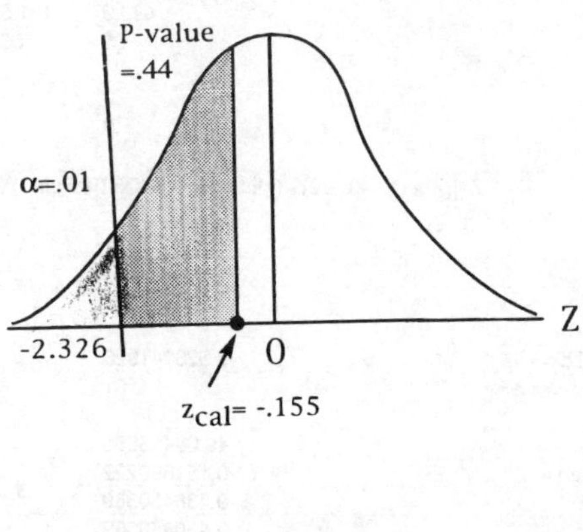

In EXCEL we can use the functions NORMSDIST or NORMDIST to obtain the cumulative probability on the left-hand side of the z-value. Type the label 'Cumulative probability =' in cell B22 and 'p-value=' in cell B23. To calculate this cumulative probability corresponding to $z_{calculated}$ = -.155 (the data-based value of the test statistic calculated above), in cell C22, type **=NORMSDIST(-.155)** or **=NORMSDIST(C21)** or **NORMDIST(C21,0,1,TRUE)** or **=NORMDIST(C5-D5,C8,C20,TRUE)**. The *p*-value for a (left) one tail test will be this cumulative probability. In cell C23, type the cell reference **=C22**. This will give the required *p*-value.

Task 6 ***Conclusion***

(a) Method A From Task 5, $z_{calculated} = -.155$

From Task 4, $z_{critical} = -2.326$

Based on the decision rule developed in Task 4, as $z_{calculated} = -.155 > z_{critical} = -2.326$, do not reject H_0: $\mu_1 - \mu_2 = 0$. In cells B25 and C25, type the text, 'Based on Method A' and 'Since Zcal = -.155 > Zcrit = -2.326, do not reject Ho'.

(b) Method B From Task 5, *p*-value = .438

Based on the decision rule given in Task 4, as the level of significance $\alpha = .01 < p$-value, do not reject H_0: $\mu_1 - \mu_2 = 0$. In cells B26 and C26, type the text, 'Based on Method B' and 'Since p-value = .438 > alpha = .01, do not reject Ho'.

Therefore, based on sample information, there is not enough evidence to conclude that male scores are lower than female scores at the 1 percent level of significance.

Task 7 Save your workbook with filename **A:EX111.XLS** on your disk and print the file. The output from Tasks 1-6 is given in Figure 11.1.

Figure 11.1 Hypothesis testing about $\mu_1 - \mu_2$ when σ_1 and σ_2 are known

	A	B	C	D	E	F
1	Hypothesis Testing about Population Mean Difference D = (Mu1 - Mu2) when D=0					
2	Population Standard Deviations Sigma1 and Sigma2 are Known					
3					Male SAT	Female SAT
4	Data Summary:	Sample size n =	6	8	290	260
5		Sample mean Xbar =	423.9	431.5	418	330
6		Population SD sigma =	95	85	507	445
7					469	520
8	Step 1	Ho: (Mu1 - Mu2) =	0		510	490
9		HA :(Mu1 - Mu2) <	0		349.4	535
10						420
11	Step 2	Test statistic:	Z=[(Xbar1 - Xbar2) - (Mu1 - Mu2)]/SQRT[(sigma1^2/n1)+((sigma2^2/n2)]			452
12						
13	Step 3	Level of significance =	0.01			
14						
15	Step 4	Critical value z(.01) =	-2.326341928			
16		Decision rule:				
17		Method A: Reject Ho if Zcal <	-2.326341928			
18		Method B: Reject Ho if p-value <	0.01			
19						
20	Step 5	SE of sample mean =	49.06415868			
21		Test statistic Z-calculated =	-0.154899222			
22		Cumulative probability =	0.438450369			
23		p-value =	0.438450369			
24						
25	Step 6	Based on Method A	Since Zcal = -.155 > Zcrit = -2.326, do not reject Ho			
26		Based on Method B	Since p-val = .438> alpha = .01, do not reject Ho			
27						
28						
29						

Method 2

Under Method 2, Tasks 1-3 and Task 6 are exactly the same as under Method 1. In Method 2, to perform Tasks 4 and 5, we use the *Data Analysis*... option under *Tools* on the *Menubar* (see Figure 3.1 and footnote 1, Chapter 3). This gives the critical values (Task 4) and the data-based value of the test statistic as well as the corresponding *p*-value (Task 5) automatically without the need for any formulas.

Tasks 1-3 Same as Method 1.

Tasks 4-5 First, click **Tools** on the *Menubar* and select the option **Data Analysis...**. From the *Analysis Tools dialogue box*, since the population variances are known and the populations are normally distributed (Case 1, Section 11.1), select **z-Test: Two Sample for Means.** Now a *z-Test: Two sample for Means dialogue box* will appear on the screen (see Figure 11.2). The **Input** section will ask for 7 inputs, namely, *Variable 1 Range* (enter **E3:E9**); *Variable 2 Range* (enter **F3:F11**); *Hypothesised Mean Difference* (enter the value of D which is **0**); *Variable 1 Variance* (enter **9025**)[1]; *Variable 2 Variance* (enter **7225**)[1]; *Label* (click on the little square as the labels in cell E3 and F3 are included in the data range); *Alpha* (enter **.01**). Under **Output options,** click on the little circle against *Output Range* and enter cell number **A30**. Finally, click **OK**. Figure 11.3 presents the output. The required output for Tasks 4 and 5 (critical value, value of the test statistic and the *p*-value) can be found in Figure 11.3.

Figure 11.2 z-Test: Two Sample for Means dialogue box

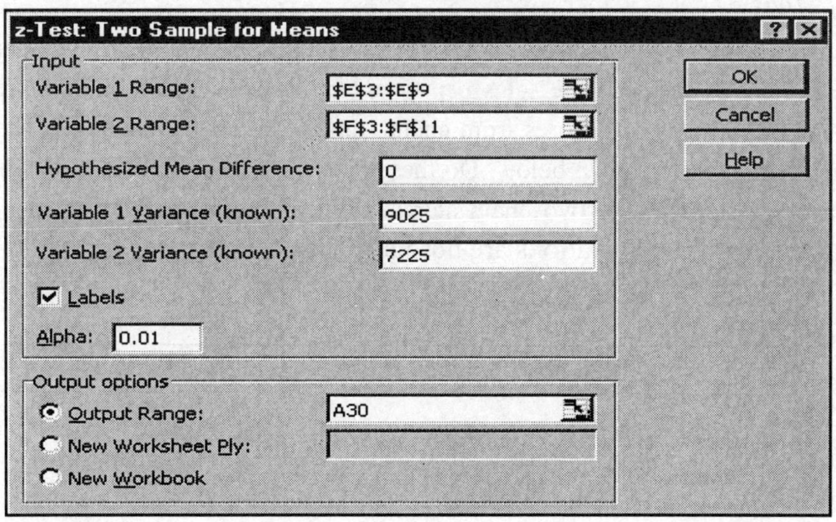

1 It should be noted that for the variances (known), EXCEL does not accept cell references. The values should be typed-in. If standard deviations are given, calculate the variances elsewhere and enter the variance values.

Figure 11.3 z-Test: Two Sample for Means output

	A	B	C
30	z-Test: Two Sample for Means		
31			
32		*Male SAT*	*Female SAT*
33	Mean	423.9	431.5
34	Known Variance	9025	7225
35	Observations	6	8
36	Hypothesized Mean Difference	0	
37	z	-0.154899222	
38	P(Z<=z) one-tail	0.438450369	
39	z Critical one-tail	2.326341928	
40	P(Z<=z) two-tail	0.219225184	
41	z Critical two-tail	2.575834515	

As you can see from the output, the critical value, data-based value of the test statistic and the corresponding *p*-value of Tasks 4 and 5 from both methods are the same. Consequently, Tasks 6 and 7 will be the same for both methods.

11.3 Example 11.2: Testing hypotheses about the difference between two population means D = (μ_1 - μ_2) when D = 0, and σ_1^2 and σ_2^2 are unknown

(Ref: Based on Example 11.3, p.129, Australian Business Statistics, Study Guide, Selvanathan et al., 1995)

A statistician wants to compare the relative success of two large department store chains in Brisbane. She decides to measure the sales per square metre space each chain store uses. She takes random samples of 5 stores from chain 1 and 8 stores from chain 2. The gross sales per square metre are shown below. Do these data provide sufficient evidence to indicate that average sales in the two chain stores differ at the 10 percent level of significance? Assume that the populations are normally distributed.

Sales($/square metre)

Chain 1	Chain 2
65.50	82.50
72.00	63.50
93.00	68.00
88.50	70.00
82.60	66.50
	75.00
	69.00
	71.00

Perform the following tasks as intermediate steps to reach your conclusion.

Tasks 1-6 Perform the same Tasks 1-6 as in Example 11.1. Use EXCEL to perform Tasks 4 and 5. Include in your EXCEL file information regarding the other tasks.

Task 7 Save your workbook as 'A:EX112.XLS' on your disk and print the file.

Solution to Example 11.2

Let X_1 and X_2 be the sales per square metre space of a chain 1 store and a chain 2 store; μ_1 and μ_2 be the population mean sales per square metre; and σ_1 and σ_2 be the population standard deviations, respectively. In cell A1, type the title 'Hypothesis Testing about Population Mean Difference D=(Mu1 – Mu2) when D=0' and center it across cells A1 to D1. Type second title 'Population Standard Deviations Sigma1 and Sigma2 are Unknown' in cell A2 and center it across cells A2 to D2.

Data summary (given)

Sample sizes $n_1 = 5$, $n_2 = 8$; sample means $\overline{X}_1$ and $\overline{X}_2$ are to be calculated; population standard deviations σ_1 and σ_2 are both unknown and estimated by the sample standard deviations s_1 and s_2, respectively, which are to be calculated; and significance level α = .10.

In cells C4 and D4, type column headings 'Sample 1' and 'Sample 2'. In cells C5 to C9, type the observations for sample 1 and, in cells D5 to D12, type the observations for sample 2. In cells B14 to B16, type the labels 'Sample size n =', 'Sample mean Xbar =' and 'Sample variance S^2=' To input the given information, in cells C14 and D14, type the corresponding values **5** and **8.** To obtain the two sample means, in cells C15 and D15, type **=AVERAGE(C5:C9)** and **=AVERAGE(D5:D12)**. To obtain the two sample variances, in cells C16 and D16, type **=VAR(C5:C9)** and **=VAR(D5:D12)**.

Note: In this example, as we have access to raw data, we can use either Method 1 or Method 2 (described below) to perform Tasks 4 and 5.

Method 1

Task 1 To test whether the average sales in the two chain stores differ, we set up the null and the alternative hypotheses as follows:

$$H_0: \quad \mu_1 - \mu_2 = 0$$
$$H_A: \quad \mu_1 - \mu_2 \neq 0$$

As the alternative hypothesis contains the 'not equal' sign, the rejection region of the test is in both tails and this is a two tail test.

In cells B18 and B19, type the labels 'Ho: (Mu1 - Mu2) =' and 'HA:(Mu1 – Mu2) not equal to' and, in cells C18 and C19, enter the values, **0** and **0**.

Task 2 As the population standard deviations σ_1 and σ_2 are *unknown* and sample sizes are small, assuming that the populations are normally distributed, to test a hypothesis about the difference between two population means ($\mu_1 - \mu_2$), we use the test statistic, (see Case 2, Section 11.1)

$$t = \frac{(\overline{X}_1 - \overline{X}_2) - (\mu_1 - \mu_2)}{\sqrt{\frac{s_1^2}{n_1} + \frac{s_2^2}{n_2}}} \sim t_\nu$$

That is, we use the t-distribution with ν degrees of freedom, where

$$\nu = \frac{\left[\frac{s_1^2}{n_1} + \frac{s_2^2}{n_2}\right]^2}{\frac{(s_1^2/n_1)^2}{n_1 - 1} + \frac{(s_2^2/n_2)^2}{n_2 - 1}}$$

In cells B21 and C21, type the labels 'Test Statistic:' and 't = (Xbar1 – Xbar2) – (Mu1 – Mu2)/SQRT[(S1^2/n1)+(S2^2/n2)]'. In cell B22, type the label 'Degrees of freedom ='. To calculate the degrees of freedom, ν, in cell C22, type the formula **=(((C16/C14)+(D16/D14))^2/(((C16/C14)^2/(C14-1))+((D16/D14)^2/(D14-1)))).**

Task 3 In cell B25, type the label, 'Level of significance =' and, in cell C25, type the given value **0.10**.

Task 4 ***Calculating the critical values***

As this is a two tail test (see diagram below), for a test with 10 percent level of significance,

$$\text{Critical value } t_{critical} = \pm t_{\alpha,\nu} = \pm t_{.05,\nu}$$

where the degrees of freedom ν is the value calculated in cell C22. This critical value is based on a t-distribution and can be obtained using the EXCEL function TINV. We follow the same procedure as in Task 4 of Example 10.4 (Section 10.6). In cell B27, type 'Critical values -t(.05,d.f.) and t(.05,d.f.)='. To obtain the critical values, in cell D27, type =**TINV(.1,C22)** or =**TINV(C25,C22)**. The critical value ($t_{critical}$) of 2.015 will appear in cell D27. In cell C27, type =**-D27**.

The decision rule

Therefore the decision rule for the test is

Method A: **Reject H_0 if $|t_{calculated}| > t_{critical} = 2.015$**

That is, Reject H_0 if $t_{calculated} < -t_{critical} = -2.015$

or $t_{calculated} > t_{critical} = 2.015$

Method B: **Reject H_0 if *p*-value $< \alpha = .10$**

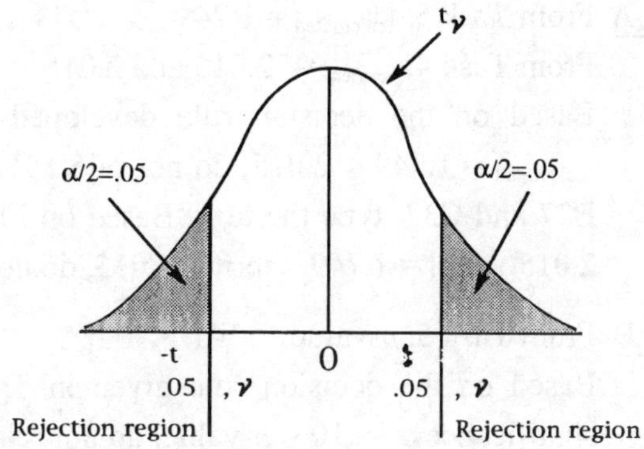

In cells B28 to B30, type the labels 'Decision Rule: ', 'Method A: Reject Ho if t-cal < or >' and 'Method B: Reject Ho if p-value <'. In cells C29, D29 and C30, type the cell references **=C27** , **=D27** and **=C25**.

Task 5 ***Calculating the data-based value of the test statistic ($t_{calculated}$)***

To calculate the data-based value of the test statistic defined in Task 2, first we calculate the denominator which is an estimator for the standard error of ($\overline{X}_1$ - $\overline{X}_2$) and then calculate the data-based value of the test statistic ($t_{calculated}$). In cells B32 and B33, type the labels 'SE of sample mean =' and 'Test statistic t-calculated ='. To obtain the estimate for the standard error of the sample mean, in cell C32, type the formula **=SQRT((C16/C14)+(D16/D14))**.

To calculate data-based value of the test statistic, $t_{calculated}$, in cell C33, type the formula **=(C15-D15-C18)/C32**. This will give the value (=1.749) for the test statistic in cell C33.

Calculating the p-value

For a two tail t-test, the *p*-value = 2P[t > ABS($t_{calculated}$)], where ABS($t_{calculated}$) is the absolute value of the data-based value of the test statistic calculated above.

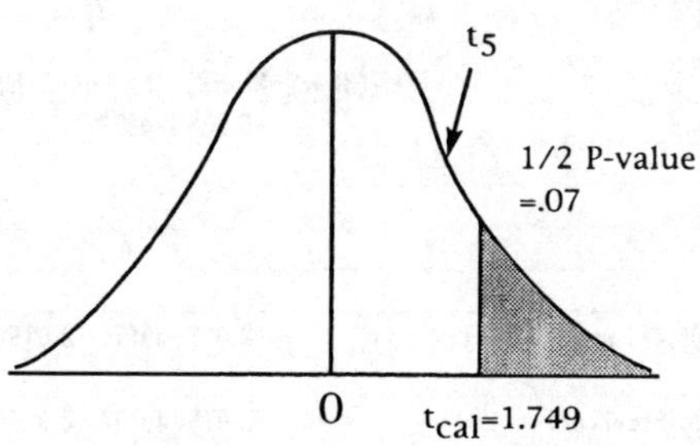

As this is a two tail test and uses the t-distribution, we use TDIST to calculate the two tail probability. In cells B34 and B35, type the labels 'Two tail probability' and 'p-value='. In cell C34, type the formula **=TDIST(ABS(C33),C22,2)**. This will give the two tail probability value of .1407 in cell C34. In cell C35, type the formula **=C34**, to obtain the *p*-value.

Task 6 ***Conclusion***

(a) Method A From Task 5, $t_{calculated} = 1.749$

From Task 4, $t_{critical} = -2.015$ and 2.015

Based on the decision rule developed in Task 4, $-2.015 < t_{calculated} = 1.749 < 2.015$, do not reject H_0: $\mu_1 - \mu_2 = 0$. In cells B37 and C37, type the text 'Based on Method A' and 'Since -2.015 < tcal = 1.749 < tcrit = 2.015, do not reject Ho'.

(b) Method B From Task 5, *p*-value = .141

Based on the decision rule given in Task 4, as the level of significance $\alpha = .10 < p$-value, do not reject H_0: $\mu_1 - \mu_2 = 0$. In cells B38 and C38, type the text, 'Based on Method B' and 'Since p-value = .141 > alpha = .10, do not reject Ho'.

Task 7 Save your workbook as **A:EX112.XLS** on your disk and print the file. The output from Tasks 1-6 is given in Figure 11.4.

Figure 11.4 Hypothesis testing about μ_1 - μ_2 when σ_1 and σ_2 are unknown

	A	B	C	D	E	F	G	H
1	Hypothesis Testing about Population Mean Difference D = (Mu1 - Mu2) when D=0							
2		Population Standard Deviations Sigma1 and Sigma2 are Unknown						
3								
4			Sample 1	Sample 2				
5			65.5	82.5				
6			72	63.5				
7			93	68				
8			88.5	70				
9			82.6	66.5				
10				75				
11				69				
12				71				
13								
14	Data Summar	Sample size n:	5	8				
15		Sample mean Xbar:	80.32	70.6875				
16		Sample variance S^2:	130.437	33.9955				
17								
18	Step 1	Ho: (Mu1 - Mu2):	0					
19		HA: (Mu1 - Mu2) not equal to	0					
20								
21	Step 2	Test statistic:	t=[(Xbar1 - Xbar2) - (Mu1-Mu2)]/SQRT[(S1^2/n1)+(S2^2)/n2]					
22		Degrees of freedom:	5.328484022					
23								
24								
25	Step 3	Level of significance:	0.1					
26								
27	Step 4	Critical values -t(.05,df) and t(.05,df):	-2.015049176	2.01505				
28		Decision rule:						
29		Method A: Reject Ho if tcal < or >	-2.015049176	2.01505				
30		Method B: Reject Ho if p-value <	0.1					
31								
32	Step 5	SE of sample mean:	5.507889066					
33		Test statistic t-calculated:	1.748855121					
34		Two Tail probability:	0.140729823					
35		p-value:	0.140729823					
36								
37	Step 6	Based on Method A	Since -2.015 < tcal = 1.749 < tcrit = 2.015, do not reject Ho					
38		Based on Method B	Since p-value = .141 > alpha = .10, do not reject Ho					

Method 2

Under Method 2, Tasks 1-3 and Task 6 are exactly the same as under Method 1. In Method 2, to perform Tasks 4 and 5, we use the *Data Analysis*... option under *Tools* on the *Menubar* (see Figure 3.1 and footnote 1, Chapter 3). This gives the critical values (Task 4) and the data-based value of the test statistic as well as the corresponding *p*-value (Task 5) automatically without the need for any formulas.

Tasks 1-3 Same as Method 1.

Tasks 4-5 First, click **Tools** on the *Menubar* and select the option **Data Analysis....** From the *Analysis Tools dialogue box*, since the population variances are unknown and sample sizes are small, assuming that the populations are normally distributed (Case 2, Section 11.1), select **t-Test: Two Sample Assuming Unequal Variances**. Now a *t-Test: Two Sample Assuming Unequal Variances dialogue box* will appear on the screen (see Figure 11.5). The **Input** section will ask for 5 inputs, namely, *Variable 1 Range* (enter **C4:C9**); *Variable 2 Range* (enter **D4:D12**); *Hypothesized Mean Difference* (enter the value of D which is **0**); *Label* (click on the little square as the labels in cell C4 and D4 are included in the data range); *Alpha* (enter **.10**). Under **Output options,** click on the little circle against *Output Range* and enter cell number **A40**. Finally, click **OK**. Figure 11.6 presents the output. The required output for Tasks 4 and 5 (critical value, value of the test statistic and the *p*-value) can be found in Figure 11.6.

Figure 11.5 t-Test: Two-Sample Assuming Unequal Variances dialogue box

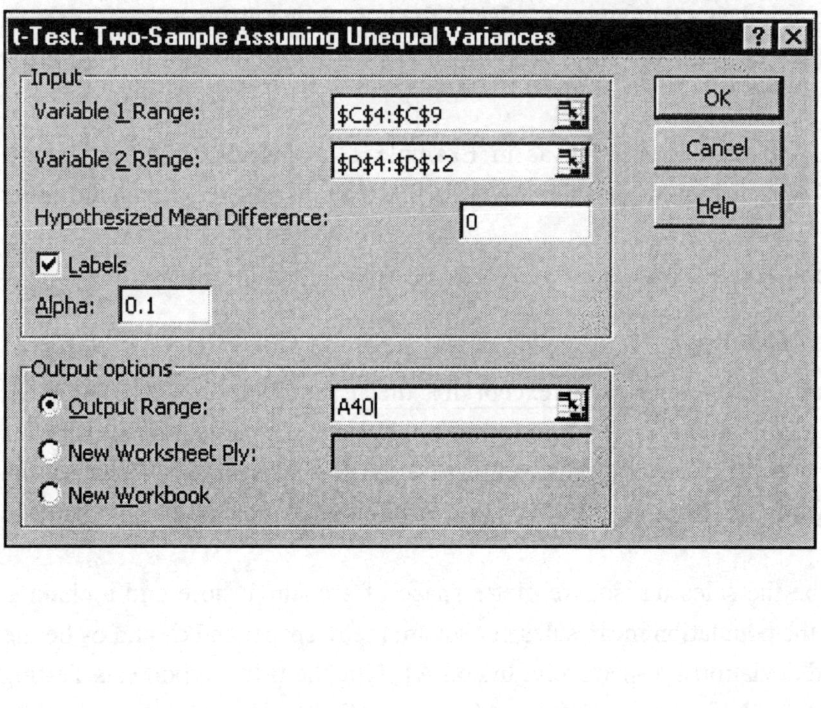

Figure 11.6 t-Test: Two-Sample Assuming Unequal Variances output

	A	B	C
40	t-Test: Two-Sample Assuming Unequal Variances		
41			
42		*sample1*	*sample2*
43	Mean	80.32	70.6875
44	Variance	130.437	33.99553571
45	Observations	5	8
46	Hypothesized Mean Difference	0	
47	df	5	
48	t Stat	1.7488551	
49	P(T<=t) one-tail	0.0703649	
50	t Critical one-tail	1.4758848	
51	P(T<=t) two-tail	0.1407298	
52	t Critical two-tail	2.0150492	

11.4 Example 11.3: Testing hypotheses about the difference between two population means D = (μ_1 - μ_2) when D = 0 and σ_1^2 and σ_2^2 are unknown but equal

(Ref: Based on Example 11.3, p. 129, Australian Business Statistics, Study Guide, Selvanathan et al., 1995)

Consider Example 11.2 (Section 11.3 above). Assuming that the two populations are normally distributed and the population variances are equal, do these data provide sufficient evidence to indicate that average sales in the two chains differ at the 10 percent level of significance? Perform the following tasks as intermediate steps to reach your conclusion.

Tasks 1-6 Perform the same Tasks 1-6 as in Example 11.1. Use EXCEL to perform Tasks 4 and 5. Include in your EXCEL file information regarding the other tasks.

Task 7 Save your workbook as 'A:EX113.XLS' on your disk and print the file.

Solution to Example 11.3

Examples 11.2 and 11.3 are the same except for the assumption that the unknown variances are equal in Example 11.3. In performing the tasks, the calculations differ for the estimator of the population variances, standard error of the difference between the sample means, degrees of freedom, critical values, data-based value of the test statistic and *p*-value.

Let X_1 and X_2 be the sales per square metre space of a chain 1 store and a chain 2 store; μ_1 and μ_2 be the population mean sales per square metre space; and σ_1 and σ_2 be the population standard deviations, respectively. In cell A1, type the title 'Hypothesis Testing about Population Mean Difference D=(Mu1 – Mu2) when D=0' and center it across cells

A1 to D1. Type second title 'Population Standard Deviations Sigma1 and Sigma2 are Unknown but Equal' in cell A2 and center it across cells A2 to D2.

Data summary (given)

Sample sizes $n_1 = 5$, $n_2 = 8$; sample means $\overline{X}_1$ and $\overline{X}_2$ are to be calculated; population variances σ_1^2 and σ_2^2 are both unknown but equal and estimated by the pooled variance which is the weighted average of the sample variances s_1^2 and s_2^2, and is to be calculated; and significance level $\alpha = .10$.

In cells C4 and D4, type column headings 'Sample 1' and 'Sample 2'. In cells C5 to C9, type the observations for sample 1 and, in cells D5 to D12, type the observations for sample 2. In cells B14 to B16, type the labels 'Sample size n =', 'Sample mean Xbar =' and 'Sample variance S^2=' To input the given information, in cells C14 and D14, type the corresponding values **5** and **8.** To obtain the two sample means, in cells C15 and D15, type **=AVERAGE(C5:C9)** and **=AVERAGE(D5:D12)**. To obtain the two sample variance, in cells C16 and D16, type **=VAR(C5:C9)** and **=VAR(D5:D12)**.

Note: In this example, as we have access to raw data, we can use either Method 1 or Method 2 (described below) to perform Tasks 4 and 5.

Method 1

Task 1 To test whether the average sales in the two chain stores differ, we set up the null and the alternative hypotheses as follows:

H_0: $\mu_1 - \mu_2 = 0$

H_A: $\mu_1 - \mu_2 \neq 0$

As the alternative hypothesis contains the 'not equal' sign, the rejection region of the test is in both tails and this is a two-tail test.

In cells B18 and B19, type the labels 'Ho: (Mu1 - Mu2) =' and 'HA:(Mu1 – Mu2) not equal to' and, in cells C18 and C19, enter the values, **0** and **0.**

Task 2 As the populations are normally distributed, the population standard deviations σ_1 and σ_2 are *unknown but equal* and sample sizes are small, to test a hypothesis about the difference between two population means ($\mu_1 - \mu_2$), we use the test statistic (see Case 3, Section 11.1)

$$t = \frac{(\overline{X}_1 - \overline{X}_2) - (\mu_1 - \mu_2)}{\sqrt{\left(\frac{s_P^2}{n_1} + \frac{s_P^2}{n_2}\right)}} \sim t_{n_1+n_2-2}$$

where

$$s_P^2 = \frac{(n_1 - 1)s_1^2 + (n_2 - 1)s_2^2}{n_1 + n_2 - 2}$$

That is, we use the t-distribution with $n_1+n_2-2 = 11$ degrees of freedom.

In cells B21 and C21, type the labels 'Test Statistic:' and 't = (Xbar1 – Xbar2) – (Mu1 – Mu2)/SQRT[(Sp^2/n1)+(Sp^2/n2)]'. In cell B22, type the label 'Degrees of freedom ='. To calculate the degrees of freedom, in cell C22, type the formula **=C14+D14-2.** In cell B23, type the label 'Pooled variance Sp^2='. To calculate the pooled variance, in cell C23, type the formula **=((C14-1)*C16+(D14-1)*D16)/C22.**

Task 3 In cell B25, type the label 'Level of significance =' and, in cell C25, type the given value **0.10**.

Task 4 ***Calculating the critical values***

As this is a two tail test (see diagram below), for a test with 10 percent level of significance,

$$\text{Critical value } t_{critical} = \pm t_{\alpha/2,n_1+n_2-2} = \pm t_{.05,11}$$

This value is based on a t-distribution and can be obtained using the EXCEL function TINV. We follow the same procedure as in Task 4 of Example 10.4 (Section 10.6). In cell B27, type 'Critical value -t(.05,df) and t(.05,df) ='. To obtain the critical value, in cell D27, type the formula = **TINV(.1,11)** or **=TINV(C25,C22)**. The critical value ($t_{critical}$) of 1.795884 will appear in cell D27. In cell C27, type **=-D27**.

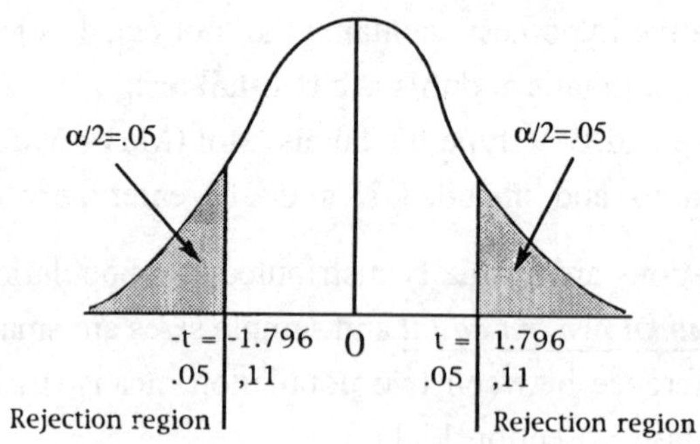

The decision rule

Therefore the decision rule for the test is

Method A: **Reject H_0 if $|t_{calculated}| > t_{critical} = 1.796$**
That is, Reject H_0 if $t_{calculated} < -t_{critical} = -1.796$
or $t_{calculated} > t_{critical} = 1.796$

Method B: **Reject H_0 if *p*-value $< \alpha = .10$**

In cells B28 to B30, type the labels 'Decision Rule: ', 'Method A: Reject Ho if t-cal < or >' and 'Method B: Reject Ho if p-value <'. In cells C29, D29 and C30, type the cell references **=C27** , **=D27** and **=C25**.

Task 5 ***Calculating the data-based value of the test statistic ($t_{calculated}$)***

To calculate the data-based value of the test statistic defined in Task 2, first we calculate the denominator which is an estimator for the standard error of ($\overline{X}_1$ - $\overline{X}_2$) and then calculate the data-based value of the test statistic ($t_{calculated}$). In cells B32 and B33, type the labels 'SE of sample mean =' and 'Test statistic t-calculated ='. To obtain the estimate for the standard error of the sample mean, in cell C32, type the formula **=SQRT((C23/C14)+(C23/D14))**.

To calculate data-based value of the test statistic, $t_{calculated}$, in cell C33, type the formula **=(C15-D15-C18)/C32**. This will give the value (=2.033) for the test statistic in cell C33.

Calculating the p-value

For a two tail t-test, the *p*-value = 2P[t > ABS($t_{calculated}$)], where ABS($t_{calculated}$) is the absolute value of the data-based value of the test statistic calculated above.

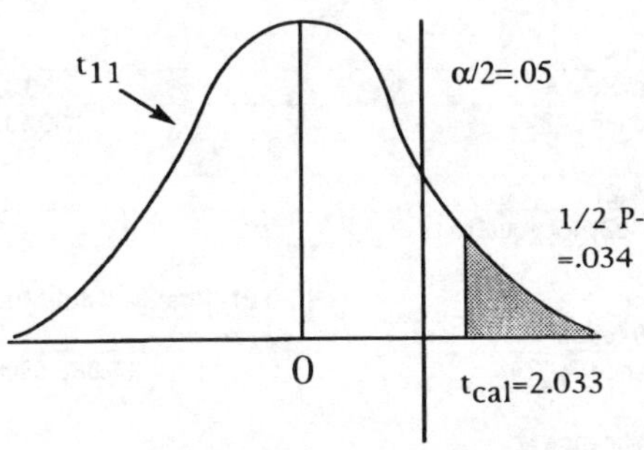

As this is a two tail test and uses the t-distribution, we use TDIST to calculate the two tail probability. In cells B34 and B35, type the labels 'Two tail probability' and 'p-value='. In cell C34, type the formula **=TDIST(ABS(C33),C22,2)**. This will give the two tail probability value of .067 in cell C34. In cell C35, type the formula **=C34**, to obtain the *p*-value.

Task 6 ***Conclusion***

(a) Method A From Task 5, $t_{calculated}$ = 2.033

From Task 4, $t_{critical}$ = -1.796 and 1.796

Based on the decision rule developed in Task 4, $t_{calculated}$ = 2.033 > $t_{critical}$ = 1.796, reject H_0: $\mu_1 - \mu_2 = 0$ in favour of the alternative H_A: $\mu_1 - \mu_2 \neq 0$. In cells B37 and C37, type the text, 'Based on Method A' and 'Since tcal = 2.033 > tcrit = 1.796, reject Ho'.

(b) Method B From Task 5, p-value = .067

Based on the decision rule given in Task 4, as the level of significance $\alpha = .10 < p$-value, reject H_0: $\mu_1 - \mu_2 = 0$ in favour of the alternative H_A: $\mu_1 - \mu_2 \neq 0$. In cells B38 and C38, type the text, 'Based on Method B' and 'Since p-value = .067 < alpha = .10, reject Ho'.

Task 7 Save your workbook as **A:EX113.XLS** on your disk and print the file. The output from Tasks 1-6 is given in Figure 11.7.

Figure 11.7 Hypothesis testing about $\mu_1 - \mu_2$ when σ_1 and σ_2 are unknown but assumed equal

	A	B	C	D
1		Hypothesis Testing about Population Mean Difference D = (Mu1- Mu2) when D=0		
2		Population Standard Deviations Sigma1 and Sigma2 are Unknown but are Equal		
3				
4			Sample 1	Sample 2
5			65.5	82.5
6			72	63.5
7			93	68
8			88.5	70
9			82.6	66.5
10				75
11				69
12				71
13				
14	Data Summary:	Sample size n =	5	8
15		Sample mean Xbar =	80.32	70.6875
16		Sample variance S^2 =	130.437	33.99553571
17				
18	Step 1	Ho: (Mu1 - Mu2) =	0	
19		HA: (Mu1 - Mu2) not equal to	0	
20				
21	Step 2	Test statistic:	t=[(Xbar1 - Xbar2) - (Mu1-Mu2)]/SQRT[(Sp**2/n1)+(Sp**2)/n2]	
22		Degrees of freedom =	11	
23		Pooled variance Sp^2 =	69.06515909	
24				
25	Step 3	Level of significance =	0.1	
26				
27	Step 4	Critical values -t(.10,df) and t(.10,df)=	-1.7958837	1.795883691
28		Decision rule:		
29		Method A: Reject Ho if tcal < or >	-1.7958837	1.795883691
30		Method B: Reject Ho if p-value <	0.1	
31				
32	Step 5	SE of sample mean =	4.73773962	
33		Test statistic t-calculated =	2.033142548	
34		Two Tail probability =	0.066888478	
35		p-value =	0.066888478	
36				
37	Step 6	Based on Method A	Since tcal =2.033 > tcrit = 1.796, reject Ho	
38		Based on Method B	Since p-val = .067 < alpha = .10, reject Ho	
39				

Method 2

As in Example 11.2, under Method 2, Tasks 1-3 and Task 6 are exactly the same as under Method 1. In Method 2, to perform Tasks 4 and 5, we use the **Data Analysis...** option under *Tools* on the *Menubar* (see Figure 3.1 and footnote 1, Chapter 3). This gives the critical values (Task 4) and the data-based value of the test statistic as well as the corresponding p-value (Task 5) automatically without the need for any formulas.

Tasks 1-3 Same as Method 1.

Tasks 4-5 First, click **Tools** on the *Menubar* and select the option **Data Analysis....** From the *Analysis Tools dialogue box*, since the population variances are unknown but equal and sample sizes are small, assuming that the populations are normally distributed (Case 3, Section 11.1), select **t-Test: Two Sample Assuming Equal Variances**. Now a *t-Test: Two Sample Assuming Equal Variances dialogue box* will appear on the screen (see Figure 11.8). The **Input** section will ask for 5 inputs, namely, *Variable 1 Range* (enter **C4:C9**); *Variable 2 Range* (enter **D4:D12**); *Hypothesized Mean Difference* (enter the value of D which is **0**); *Label* (click on the little square as the labels in cell C4 and D4 are included in the data range); *Alpha* (enter **.10**). Under **Output options,** click on the little circle against *Output Range* and enter cell number **A40**. Finally, click **OK**. Figure 11.9 presents the output. The required output for Tasks 4 and 5 (critical value, value of the test statistic and the *p*-value) can be found in Figure 11.9.

Figure 11.8 t-Test: Two-Sample Assuming Equal Variances dialogue box

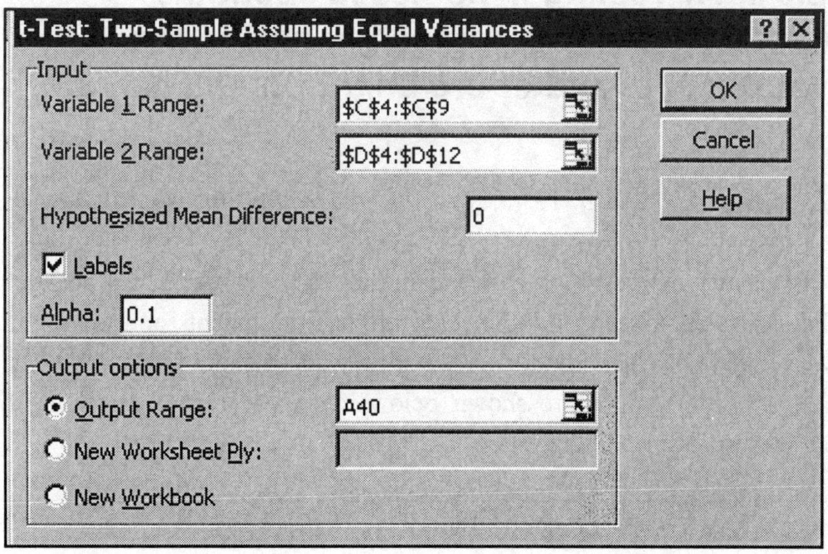

Figure 11.9 t-Test: Two-Sample Assuming Equal Variances output

	A	B	C
40	t-Test: Two-Sample Assuming Equal Variances		
41			
42		*Variable 1*	*Variable 2*
43	Mean	80.32	70.6875
44	Variance	130.437	33.99553571
45	Observations	5	8
46	Pooled Variance	69.06515909	
47	Hypothesized Mean Difference	0	
48	df	11	
49	t Stat	2.033142548	
50	P(T<=t) one-tail	0.033444239	
51	t Critical one-tail	1.363430329	
52	P(T<=t) two-tail	0.066888478	
53	t Critical two-tail	1.795883691	

11.5 Example 11.4: Testing hypotheses about the difference between two population means D = (μ_1 - μ_2) when D = 0, σ_1^2 and σ_2^2 are unknown and n_1 > 30 and n_2 > 30

(Ref: Based on Exercises 11.12 and 11.13, p. 309, *Australian Business Statistics*, *Selvanathan et al., 1994)*

Do students at three-year universities work harder than those at two-year TAFE college? To help answer this question, 47 randomly selected university students and 36 TAFE college students were asked how many hours per week they spent doing homework. The means and variances for both groups are shown below. Do these results allow us to answer our opening question affirmatively? (Use $\alpha = 0.10$.)

University students	TAFE college students
$\overline{X}_1 = 18.6$	$\overline{X}_2 = 14.7$
$s_1^2 = 22.4$	$s_2^2 = 20.9$

Perform the following tasks as intermediate steps to reach your conclusion.

Tasks 1-6 Perform the same Tasks 1-6 as in Example 11.1. Use EXCEL to perform Tasks 4 and 5. Include in your EXCEL file information regarding the other tasks.

Task 7 Save your workbook as 'A:EX114.XLS' on your disk and print the file.

Solution to Example 11.4

Let X_1 and X_2 be the number of hours per week spent doing homework by a University student and a TAFE student; μ_1 and μ_2 be the population mean number of hours; and σ_1 and σ_2 unknown population standard deviations, respectively. In cell A1, type the title 'Hypothesis Testing about Population Mean Difference D=(Mu1 – Mu2) when D=0' and center it across cells A1 to D1. Type the second title, 'Population Standard Deviations Sigma1 and Sigma2 are Unknown and n1 > 30 and n2 > 30' in cell A2 and center it across cells A2 to D2.

Data summary (given)

Sample sizes $n_1 = 47$, $n_2 = 36$; sample mean $\bar{X}_1 = 18.6$ and $\bar{X}_2 = 14.7$; sample variances $s_1^2 = 22.4$ and $s_2^2 = 20.9$; and significance level $\alpha = .10$.

In cells B4 to B6, type the labels 'Sample size n =', 'Sample mean Xbar =' and 'Sample variance S^2 ='. To input the given information, in cells C4 to C6, type the corresponding Sample 1 values **47, 18.6** and **22.4** and, in cells D4 to D6, type the corresponding Sample 2 values **36, 14.7** and **20.9.**

Note: In this example, as we do not have access to the raw data, Method 2 is not suitable. Therefore, we use Method 1 to perform the tasks.

Task 1 To test whether University students work harder than TAFE students, we set up the null and the alternative hypotheses as follows:

H_0: $\mu_1 - \mu_2 = 0$

H_A: $\mu_1 - \mu_2 > 0$

As the alternative hypothesis contains the 'greater than' sign, the rejection region of the test is in the right hand tail and this is a one tail test.

In cells B8 and B9, type the labels 'Ho: (Mu1 - Mu2) =', 'HA: (Mu1 – Mu2) > ' and in cells C8 and C9, the values, **0** and **0**.

Task 2 The population standard deviations σ_1 and σ_2 are *unknown* and sample sizes $n_1 = 47 > 30$ and $n_2 = 36 > 30$ are large. Therefore, assuming that the populations are normally distributed, to test a hypothesis about the difference between two population means ($\mu_1 - \mu_2$), we use the test statistic, (see Case 4, Section 11.1)

$$z = \frac{(\bar{X}_1 - \bar{X}_2) - (\mu_1 - \mu_2)}{\sqrt{\frac{s_1^2}{n_1} + \frac{s_2^2}{n_2}}} \sim t_{d.f.} \approx \text{Normal}(0,1)$$

That is, we use the approximate z-distribution.

In cells B11 and C11, type the labels 'Test Statistic:' and 'Z =[(Xbar1 – Xbar2)–(Mu1 – Mu2)]/SQRT[(s1^2/n1)+(s2^2/n2)]'.

Task 3 In cell B13, type the label 'Level of significance =' and, in cell C13, type the given value **0.10**.

Task 4 ***Calculating the critical value***

As this is a *(right) one tail* test, for a test with 10 percent level of significance,

$$\text{Critical value } z_{critical} = z_{\alpha} = z_{.10}$$

In cell B15, type 'Critical value z(.10)='. The critical value is on the right tail of the z-distribution with area .10 and to obtain the corresponding z-value, we use the cumulative probability .90. To obtain the critical value, in cell C15, type **=NORMSINV(.90)** or **=NORMSINV(1-C13)**. Now, the $z_{critical}$ value of 1.281551 will appear in cell C15.

The decision rule

Therefore the decision rule for the test is

Method A: **Reject H_0 if $z_{calculated} > z_{critical} = 1.281551$**

Method B: **Reject H_0 if *p*-value $< \alpha = .10$**

In cells B16 to B18, type the labels 'Decision Rule: ', 'Method A: Reject Ho if Zcal >' and 'Method B: Reject Ho if p-value <'. In cells C17 and C18, type the cell references **=C15** and **=C13**.

Task 5 ***Calculating the data-based value of the test statistic ($z_{calculated}$)***

To calculate the data-based value of the test statistic defined in Task 2, first we calculate the denominator which is an estimator for the standard error of ($\overline{X}_1$ - $\overline{X}_2$) and then calculate the data-based value of the test statistic ($z_{calculated}$). In cells B20 and B21, type the labels 'SE of sample mean =' and 'Test statistic Z-calculated ='. To obtain the estimate for the standard error of the sample mean, in cell C20, type the formula **=SQRT((C6/C4)+(D6/D4))**.

To calculate the data-based value of the test statistic, $z_{calculated}$, in cell C21, type the formula **=(C5-D5-C8)/C20**. This will give the value (=3.793) for the test statistic in cell C21.

Calculating the p-value

For a right tail z-test, the *p*-value = $P(Z > z_{calculated})$, where $z_{calculated}$ is the data-based value of the test statistic calculated above.

In EXCEL we can use the functions NORMSDIST or NORMDIST to obtain the cumulative probability on the left-hand side of the z-value. Type the label 'Cumulative probability =' in cell B22 and 'p-value=' in cell B23. To calculate this cumulative probability corresponding to $z_{calculated}$ = 3.793 (the data-based value of the test statistic calculated above), in cell C22, type **=NORMSDIST(3.793)** or **=NORMSDIST(C21)** or **=NORMDIST(3.793,0,1,TRUE)** or **=NORMDIST(C21,0,1,TRUE)**. To calculate the *p*-value, in cell C23, type the formula **=1-C22**. This will give the required *p*-value for a (right) one tail test.

Task 6 ***Conclusion***

(a) Method A From Task 5, $z_{calculated} = 3.793$

From Task 4, $z_{critical} = 1.281551$

Based on the decision rule developed in Task 4, as $z_{calculated} = 3.793 > z_{critical} = 1.282$, reject H_0: $\mu_1 - \mu_2 = 0$ in favour of the alternative H_A: $\mu_1 - \mu_2 > 0$. In cells B25 and C25, type the text, 'Based on Method A' and 'Since Zcal = 3.793 > Zcrit = 1.282, reject Ho'.

(b) Method B From Task 5, *p*-value = .00007

Based on the decision rule given in Task 4, as the level of significance $\alpha = .10 > p$-value, reject H_0: $\mu_1 - \mu_2 = 0$ in favour of the alternative H_A: $\mu_1 - \mu_2 > 0$. In cells B26 and C26, type the text, 'Based on Method B' and 'Since p-value = .00007 < alpha = .10, reject Ho'.

Task 7 Save your workbook as **A:EX114.XLS** on your disk and print the file. The output from Tasks 1-6 is given in Figure 11.10.

Figure 11.10 Hypothesis testing about $\mu_1 - \mu_2$ when σ_1 and σ_2 are unknown and n_1, $n_2 > 30$

	A	B	C	D	E	F
1	Hypothesis Testing about Population Mean Difference D = (Mu1 - Mu2) when D=0					
2	Population Standard Deviations Sigma1 and Sigma2 are Unknown and n1 > 30 and n2 > 30					
3						
4	Data Summary:	Sample size n =	47	36		
5		Sample mean Xbar =	18.6	14.7		
6		Sample variance S^2 =	22.4	20.9		
7						
8	Step 1	Ho: (Mu1 - Mu2) =	0			
9		HA :(Mu1 - Mu2) >	0			
10						
11	Step 2	Test statistic:	Z=[(Xbar1 - Xbar2) - (Mu1 - Mu2)]/SQRT[(S1^2/n1)+((S2^2/n2)]			
12						
13	Step 3	Level of significance =	0.1			
14						
15	Step 4	Critical value z(.10) =	1.281550794			
16		Decision rule:				
17		Method A: Reject Ho if Zcal >	1.281550794			
18		Method B: Reject Ho if p-value <	0.1			
19						
20	Step 5	SE of sample mean =	1.028178632			
21		Test statistic Z-calculated =	3.793115201			
22		Cumulative probability =	0.999925591			
23		p-value =	7.44093E-05			
24						
25	Step 6	Based on Method A	Since Zcal = 3.793 > Zcrit = 1.282, reject Ho			
26		Based on Method B	Since p-value = .00007 < alpha = .10, reject Ho			

11.6 Example 11.5: Testing hypotheses about the difference between two population means D = (μ_1 - μ_2) when D $\neq$ 0, σ_1^2 and σ_2^2 are unknown and n_1 > 30 and n_2 > 30

(Ref: Based on Example 11.4, p. 130, Australian Business Statistics, Study Guide, Selvanathan et al., 1995)

A statistician wants to compare the relative success of two large departments store chains in Brisbane. She decides to measure the sales per square metre space each chain store uses. She takes a random sample of 50 stores from chain 1 and 50 stores from chain 2. The summary statistics are shown below. Do these data provide sufficient evidence to indicate that the chain 1 stores' average sales per square metre is $3 more than that of the chain 2 stores at the 10 percent level of significance?

Sales ($/square metre)	
$\overline{X}_1$ = 51	$\overline{X}_2$ = .66
s_1^2 = .62	s_2^2 = .75

Perform the following tasks as intermediate steps to reach your conclusion.

Tasks 1-6 Perform the same Tasks 1-6 as in Example 11.1. Use EXCEL to perform Tasks 4 and 5. Include in your EXCEL file information regarding the other tasks.

Task 7 Save your workbook as 'A:EX115.XLS' on your disk and print the file.

Solution to Example 11.5

The main difference between Examples 11.3 and 11.5 are that in Example 11.3, D = 0 and in Example 11.5, D $\neq$ 0. Other differences are the large sample sizes in Example 11.5 and the assumption that the unknown variances are equal in Example 11.3. Examples 11.4 and 11.5 are similar (unknown variances and large sample sizes), except that in Example 11.4, we have D = 0 and in Example 11.5, we have D $\neq$ 0.

Let X_1 and X_2 be the sales per square metre space of a chain 1 store and a chain 2 store; μ_1 and μ_2 be the population mean sales per square metre space; and σ_1 and σ_2 be the unknown population standard deviations, respectively. In cell A1, type the title 'Hypothesis Testing about Population Mean Difference D=(Mu1 – Mu2) when D=3' and center it across cells A1 to D1. Type the second title 'Population Standard Deviations Sigma1 and Sigma2 are Unknown, n1 > 30 and n2 > 30' in cell A2 and center it across cells A2 to D2.

Data summary (given)

Sample sizes n_1 = 50, n_2 = 50; sample mean $\overline{X}_1$ = 75.51 and $\overline{X}_2$ = 71.66; sample variances s_1^2 = 133.62 and s_2^2 = 52.75; and significance level α = .10.

In cells B4 to B6, type the labels 'Sample size n =', 'Sample mean Xbar =' and 'Sample variance S^2='. To input the given information, in cells C4 to C6, type the corresponding Sample 1 values **50, 75.51** and **133.62** and, in cells D4 to D6, type the corresponding Sample 2 values **50, 71.66** and **52.75.**

Task 1 To test whether the average sales in the chain1 stores is $3 more than that of the chain 2 stores, we set up the null and the alternative hypotheses as follows:

H_0: $\mu_1 - \mu_2 = 3$

H_A: $\mu_1 - \mu_2 > 3$,

where D = 3. As the alternative hypothesis contains the 'greater than' sign, the rejection region is in the right hand tail and this is a one tail test.

In cells B8 and B9, type the labels 'Ho: (Mu1 - Mu2) =', 'HA: (Mu1 – Mu2) > ' and, in cells C8 and C9, enter the values, **3** and **3**.

Task 2 As the population standard deviations σ_1 and σ_2 are *unknown* and sample sizes $n_1 = 50 > 30$, $n_2 = 50 > 30$ are large, assuming that the populations are normally distributed, to test a hypothesis about the difference between two population means ($\mu_1 - \mu_2$), we use the test statistic, (see Case 4, Section 11.1)

$$z = \frac{(\overline{X}_1 - \overline{X}_2) - (\mu_1 - \mu_2)}{\sqrt{\frac{S_1^2}{n_1} + \frac{S_2^2}{n_2}}} \sim t_{d.f.} \approx Normal(0,1)$$

That is, we use the approximate z-distribution.

In cells B11 and C11, type the labels 'Test Statistic:' and 'Z = [(Xbar1 – Xbar2) – (Mu1 – Mu2)]/SQRT[(s1^2/n1)+(s2^2/n2)]'.

Task 3 In cell B13, type the label 'Level of significance =' and, in cell C13, type the given value **0.10**.

Task 4 ***Calculating the critical value(s)***

As this is a *(right) one tail* test, for a test with 10 percent level of significance,

$$\text{Critical value } z_{critical} = z_\alpha = z_{.10}$$

In cell B15, type 'Critical value z(.10)='. The critical value is on the right tail of the z-distribution with area .10 and to obtain the corresponding z-value, we use the cumulative probability .90. To obtain the critical value, in cell C15, type **=NORMSINV(.90)** or **=NORMSINV(1-C13)**. Now, the $z_{critical}$ value of 1.281551 will appear in cell C15.

The decision rule

Therefore the decision rule for the test is

Method A: **Reject H_0 if $z_{calculated} > z_{critical} = 1.281551$**

Method B: **Reject H_0 if *p*-value $< \alpha = .10$**

In cells B16 to B18, type the labels 'Decision Rule: ', 'Method A: Reject Ho if Zcal >' and 'Method B: Reject Ho if p-value <'. In cells C17 and C18, type the cell references, **=C15** and **=C13.**

Task 5 ***Calculating the data-based value of the test statistic ($z_{calculated}$)***

To calculate the data-based value of the test statistic defined in Task 2, first we calculate the denominator which is an estimator for the standard error of ($\overline{X}_1$ - $\overline{X}_2$) and then calculate the data-based value of the test statistic ($z_{calculated}$). In cells B20 and B21, type the labels 'SE of sample mean =' and 'Test statistic Z-calculated ='. To obtain the estimate for the standard error of the sample mean, in cell C20, type the formula **=SQRT((C6/C4)+(D6/D4)).**

To calculate the data-based value of the test statistic, $z_{calculated}$, in cell C21, type the formula **=(C5-D5-C8)/C20**. This will give the value (=.440) for the test statistic in cell C21.

Calculating the p-value

For a right tail z-test, the *p*-value = P(Z > $z_{calculated}$), where $z_{calculated}$ is the data-based value of the test statistic calculated above.

Type the label 'Cumulative probability =' in cell B22 and 'p-value=' in cell B23. To calculate this cumulative probability corresponding to $z_{calculated}$ = .440 (the data-based value of the test statistic calculated above), in cell C22, type **=NORMSDIST(.440)** or **=NORMSDIST(C21)** or **=NORMDIST(.440,0,1,TRUE)** or **=NORMDIST(C21,0,1,TRUE).** To calculate the *p*-value, in cell C23, type the formula **=1-C22**. This will give the required *p*-value of .330 for a (right) one tail test.

Task 6 ***Conclusion***

(a) Method A From Task 5, $z_{calculated}$ = .440

From Task 4, $z_{critical}$ = 1.281551

Based on the decision rule developed in Task 4, as $z_{calculated}$ =.440 < $z_{critical}$ = 1.281551, do not reject H_0: $\mu_1 - \mu_2 = 3$. In cells B25 and C25, type the text, 'Based on Method A' and 'Since Zcal = .440 < Zcrit = 1.282, do not reject Ho'.

(b) Method B From Task 5, *p*-value = .330

Based on the decision rule given in Task 4, as the level of significance α = .10 < *p*-value, do not reject H_0: $\mu_1 - \mu_2 = 3$. In cells B26 and C26, type the text, 'Based on Method B' and 'Since p-value = .330 > alpha = .10, do not reject Ho'.

Task 7 Save your workbook as **A:EX115.XLS** on your disk and print the file. The output from Tasks 1-6 is given in Figure 11.11.

Figure 11.11 Hypothesis testing about μ_1 - μ_2 = D when D $\neq$ 0

	A	B	C	D
1	Hypothesis Testing about Population Mean Difference D = (Mu1 - Mu2) when D =3			
2	Population Standard Deviations Sigma1 and Sigma2 are Unknown and n1 > 30 and n2 > 30			
3				
4	Data Summary:	Sample size n =	50	50
5		Sample mean Xbar =	75.51	71.66
6		Sample variance S^2 =	133.62	52.75
7				
8	Step 1	Ho: (Mu1 - Mu2) =	3	
9		HA :(Mu1 - Mu2) >	3	
10				
11	Step 2	Test statistic:	Z=[(Xbar1 - Xbar2) - (Mu1 - Mu2)]/SQRT[(s1^2/n1)+((s2^2/n2)]	
12				
13	Step 3	Level of significance =	0.1	
14				
15	Step 4	Critical value z(.10) =	1.281550794	
16		Decision rule:		
17		Method A: Reject Ho if Zcal >	1.281550794	
18		Method B: Reject Ho if p-value <	0.1	
19				
20	Step 5	SE of sample mean =	1.93064756	
21		Test statistic Z-calculated =	0.440266788	
22		Cumulative probability =	0.670128028	
23		p-value =	0.329871972	
24				
25	Step 6	Based on Method A	Since Zcal = .440 < Zcrit = 1.282, do not reject Ho	
26		Based on Method B	Since p-value = .330 > alpha = .10, do not reject Ho	

11.7 Hypothesis testing involving two populations: Qualitative data type

When making inferences about two populations where the data type is qualitative, the problem objective is to compare the population proportions. The parameter of interest is the difference between the population proportions ($p_1 - p_2$). To test hypotheses involving ($p_1 - p_2$), we follow the same six-step procedure described in Section 10.1.

In this case, the null and the alternative would take the following forms:

H_o: $p_1 - p_2 = D$

H_A: $p_1 - p_2 \neq D$ (two tail) *or*

$p_1 - p_2 < D$ (left one tail) *or*

$p_1 - p_2 > D$ (right one tail)

where D is some real value less than between –1 and +1. Usually we test for the equality of the two population proportions.

Assuming that the two samples are independent, the test statistic takes one of the following forms:

Case 1 ***H_o: $p_1 - p_2 = D$ when $D = 0$***

$$Z = \frac{(\hat{p}_1 - \hat{p}_2) - (p_1 - p_2)}{\sqrt{\frac{\hat{p}\hat{q}}{n_1} + \frac{\hat{p}\hat{q}}{n_2}}}$$

which has a standard normal distribution provided $n_1 \hat{p}_1 \geq 5$, $n_1 \hat{q}_1 \geq 5$, $n_2 \hat{p}_2 \geq 5$ and $n_2 \hat{q}_2 \geq 5$, where n_1 and n_2 are sample sizes; $\hat{q}_1 = 1 - \hat{p}_1$, and $\hat{q}_2 = 1 - \hat{p}_2$; and as $p_1 = p_2$ under H_o, we use a common proportion $\hat{p}$ as the estimate for p_1 and p_2 defined as

$$\hat{p} = \frac{\hat{X}_1 + \hat{X}_2}{n_1 + n_2} = \frac{n_1 \hat{p}_1 + n_2 \hat{p}_2}{n_1 + n_2}$$

where $\hat{X}_1 = n_1 \hat{p}_1$ and $\hat{X}_2 = n_2 \hat{p}_2$.

Case 2 ***H_o: $p_1 - p_2 = D$ when $D \neq 0$***

$$Z = \frac{(\hat{p}_1 - \hat{p}_2) - (p_1 - p_2)}{\sqrt{\frac{\hat{p}_1\hat{q}_1}{n_1} + \frac{\hat{p}_2\hat{q}_2}{n_2}}}$$

which has a standard normal distribution provided $n_1 \hat{p}_1 \geq 5$, $n_1 \hat{q}_1 \geq 5$, $n_2 \hat{p}_2 \geq 5$ and $n_2 \hat{q}_2 \geq 5$, where n_1 *and* n_2 are sample sizes, $\hat{q}_1 = 1 - \hat{p}_1$ and $\hat{q}_2 = 1 - \hat{p}_2$.

11.8 Example 11.6: Testing hypotheses about the difference between two population proportions D = (p_1 - p_2) when D = 0

(Ref: Based on Exercise 11.40, p. 326, Australian Business Statistics, Selvanathan et al, 1994.)

In a public opinion survey, 60 out of a sample of 100 high-income voters and 40 out of a sample of 75 low-income voters supported the introduction of a GST. Can we conclude at the 5% level of significance that the proportion of voters favouring a GST differs between high and low income voters? Perform the following tasks as intermediate steps to reach your conclusion.

Tasks 1-6 Perform the same Tasks 1-6 as in Example 11.1. Use EXCEL to perform Tasks 4 and 5. Include in your EXCEL file information regarding the other tasks.

Task 7 Save your workbook as 'A:EX116.XLS' on your disk and print the file.

Solution to Example 11.6

Let X_1 be the number of high-income voters and X_2 be the number of low-income voters in the population who supported the introduction of a GST. Then $p_1 = X_1/N$ and $p_2 = X_2/N$ are the proportions in the population, who supported the introduction of a GST, where N is the population size.

Let $\hat{X}_1$ be the number of high-income voters and $\hat{X}_2$ be the number of low-income voters in the samples who supported the introduction of GST. Then $\hat{p}_1 = \hat{X}_1/n_1$ and $\hat{p}_2 = \hat{X}_2/n_2$ are the sample proportions, who supported the introduction of GST, where n_1 and n_2 are the sample sizes. In cell A1, type the title 'Hypothesis Testing about Population Proportion Difference D=(p1-p2) when D=0' and centre across cells A1 to D1.

Data summary (given)

Sample sizes $n_1 = 100$, $n_2 = 75$; $\hat{X}_1 = 60$, $\hat{X}_2 = 40$; and significance level $\alpha = .05$. In cells C3 and D3, type column headings 'Sample1' and 'Sample 2'. In cells B4 to B6, type the labels 'Sample size n =', 'Sample Xhat =' and 'Sample proportion phat='. To input the given information, in cells C4 and C5, type the corresponding values **100** and **60**; and, in cells D4 and D5, type the corresponding values **75** and **40.** In cell C6 and D6, type the formulas **=C5/C4** and **=D5/D4,** respectively, to obtain the $\hat{p}_1$ and $\hat{p}_2$ values.

Task 1 We want to test whether the proportion of voters favouring a GST differs between high and low income voters. Therefore, the null and the alternative hypotheses are

$$H_0: \quad p_1 - p_2 = 0$$
$$H_A: \quad p_1 - p_2 \neq 0$$

and D = 0 under H_0. As the alternative hypothesis contains the 'not equal sign', i.e. the rejection region is in both tails, this is a two tail test.

In cells B8 and B9, type the labels 'Ho: (p1 - p2) =', 'HA: (p1 – p2) not equal to ' and, in cells C8 and C9, type the values, **0** and **0**.

Task 2 As D = 0, to test a hypothesis about the difference between two population proportions, we use the test statistic (see Case 1, Section 11.7)

$$Z = \frac{(\hat{p}_1 - \hat{p}_2) - (p_1 - p_2)}{\sqrt{\frac{\hat{p}\hat{q}}{n_1} + \frac{\hat{p}\hat{q}}{n_2}}} \sim \text{Normal}(0,1)$$

where

$$\hat{p} = \frac{n_1\hat{p}_1 + n_2\hat{p}_2}{n_1 + n_2}$$

In cells B11 and C11, type the labels 'Test Statistic:' and 'Z = [(phat1 – phat2)–(p1-p2))]/SQRT[(phat qhat/n1) +(phat qhat/n2)]'.

Task 3 In cell B13, type the label 'Level of significance =' and, in cell C13, type the given value **0.05**.

Task 4 ***Calculating the critical values***

As this is a *two tail test*, for a test with 5 percent level of significance, we have two critical regions one on each tail of the z-distribution with tail areas of 0.025. Therefore,

Critical values $z_{critical} = \pm z_{\alpha/2} = \pm z_{.025} = -z_{.025}$ and $+z_{.025}$

In cell B15, type the label 'Critical values -z(.025) and +z(.025) ='. Click on cell C15 to store the result for $-z_{.025}$. Type the formula **=NORMSINV(.025)** or **=NORMSINV(C13/2)** to obtain this value. Click on cell D15 to store the value for $+z_{.025}$ which is the negative of $-z_{.025}$. Type the formula **=-C15**. You will see the two $z_{critical}$ values -1.959961 and 1.959961 appearing in cells C15 and D15.

The decision rule

Therefore the decision rule for the test is

Method A: **Reject H_0 if $|z_{calculated}| > z_{critical} = 1.959961$**

That is, Reject H_0 if $z_{calculated} < -z_{critical} = -1.959961$

or $z_{calculated} > z_{critical} = 1.959961$

Method B: **Reject H_0 if *p*-value $< \alpha = .05$.**

In cells B16 to B18, type the labels 'Decision Rule: ', 'Method A: Reject Ho if Zcal < or >' and 'Method B: Reject Ho if p-value <'. In cells C17, D17 and C18, type **=C15**, **=D15** and **=C13**.

Task 5 ***Calculating the data-based value of the test statistic ($z_{calculated}$)***

To calculate the data-based value of the test statistic ($z_{calculated}$) defined in Task 2, we first calculate the denominator, which is an estimator for the standard error of ($\hat{p}_1 - \hat{p}_2$) and then calculate the data-based value of the test statistic. In cells B20 to B22, type the labels 'Pooled phat =', 'SE of (phat1–phat2)=' and 'Test statistic Z-calculated ='. To calculate $\hat{p}$, which is also defined in Task 2, in cell C20, type the formula **=(C5+D5)/(C4+D4)**. To obtain the standard error of $(\hat{p}_1 - \hat{p}_2)$, in cell C21, type the formula **=SQRT((C20*(1-C20)/C4)+(C20*(1-C20)/D4))**.

To calculate the data-based value of the test statistic, in cell C22, type the formula **=(C6-D6-C8)/C21**. This will give the $z_{calculated}$ value (= .88) of the test statistic.

Calculating the p-value

For a two tail z-test, the *p*-value = 2P[Z > ABS($z_{calculated}$)], where ABS($z_{calculated}$) is the absolute value of the standardised value of the test statistic.

In cell B23 and B24, type the labels 'Cumulative probability =' and 'p-value=' and click the mouse on C23. Now we use the function **NORMSDIST** or **NORMDIST** to obtain the cumulative probability corresponding to the absolute value of $z_{calculated}$ (=.88) and then obtain the *p*-value as 2(1-cumulative probability). That is, in cell C23, type the formula **NORMSDIST(ABS(C22))** or **NORMDIST(ABS(C22),0,1,TRUE)**. To calculate the *p*-value, in cell C24, type the formula **=2*(1- C23).** The will give a value of .3778.

Task 6 ***Conclusion***

(a) Method A From Task 5, $z_{calculated}$ = .88

From Task 4, $z_{critical}$ = -1.959961 and 1.959961

Based on the decision rule developed in Task 4, as $z_{calculated}$ = .88 < $z_{critical}$ = 1.96, do not reject H_0: $p_1 - p_2 = 0$. In cells B26 and C26, type the text, 'Based on Method A' and 'Since Zcal = .88 < Zcrit = 1.96, do not reject Ho'.

(b) Method B From Task 5, *p*-value = .378

Based on the decision rule given in Task 4, as the level of significance $\alpha = .05 <$ *p*-value, do not reject H_0: $p_1 - p_2 = 0$. In cells B27 and C27, type the text, 'Based on Method B' and 'Since p-value = .378 > alpha = .05, do not reject Ho'.

Therefore, based on the sample information, there is not enough evidence to allow us to conclude that support for the introduction of a GST differ among high-income and low-income voters, at the 5 percent level of significance.

Task 7 Save your workbook as **A:EX116.XLS** and print the file. The output from Tasks 1-6 is given in Figure 11.12.

Figure 11.12 Hypothesis testing about $p_1 - p_2$

	A	B	C	D	E	F
1	Hypothesis Testing about Population Proportion Difference D = (p1-p2) when D=0					
2						
3			Sample 1	Sample 2		
4	Data Summary:	Sample size n =	100	75		
5		Sample Xhat =	60	40		
6		Sample proportion phat =	0.6	0.5333		
7						
8	Step 1	Ho:(p1-p2) =	0			
9		HA :(p1-p2) not equal	0			
10						
11	Step 2	Test statistic	Z=[(phat1-phat2) - (p1-p2)]/SQRT[(phat qhat/n1)+(phat qhat/n2)]			
12						
13	Step 3	Level of significance =	0.05			
14						
15	Step 4	Critical values -z(.025) and z(.025) =	-1.959961082	1.96		
16		Decision rule:				
17		Method A: Reject Ho if Zcal < or >	-1.959961082	1.96		
18		Method B: Reject Ho if p-value <	0.05			
19						
20	Step 5	Pooled phat =	0.571428571			
21		SE of (phat1 - phat2) =	0.075592895			
22		Test statistic Z-calculated =	0.881917104			
23		Cumulative probability =	0.811089222			
24		p-value =	0.377821557			
25						
26	Step 6	Based on Method A	Since Zcal = .88 < Zcrit = 1.96, do not reject Ho			
27		Based on Method B	Since p-value = .378 > alpha = .05, do not reject Ho			

11.9 Example 11.7: Testing hypotheses about the difference between two population proportions D = (p_1 - p_2) when D $\neq$ 0

(Ref: Example 11.8, p. 324, Australian Business Statistics, Selvanathan et al, 1994.)

The process that is used to produce a complex component used in medical instruments typically results in defective rates in the 40% range. Recently two innovations have been developed. Innovation 1 appears to be more promising but is considerably more expensive to purchase and operate than innovation 2. After a careful analysis of the costs, management decides that it will adopt innovation 1 only if the proportion of defective components it produces is at least 8 percent smaller than that produced by innovation 2. In a random sample of 300 units produced by innovation 1, 33 were found to be defective. In a sample of 300 components produced by innovation 2, 84 were found to be defective. At the one percent significance level, can we conclude that management should adopt innovation 1? Perform the following tasks as intermediate steps to reach your conclusion.

Tasks 1-6 Perform the same Tasks 1-6 as in Example 11.1. Use EXCEL to perform Tasks 4 and 5. Include in your EXCEL file information regarding the other tasks.

Task 7 Save your workbook as 'A:EX117.XLS' on your disk and print the file.

Solution to Example 11.7

Let p_1 and p_2 be the population proportions of defective components produced by the two innovations. Let $\hat{X}_1$ and $\hat{X}_2$ be the number of defective components in the sample produced by Innovation 1 and Innovation 2, respectively. Then $\hat{p}_1 = \hat{X}_1/n_1$ and $\hat{p}_2 = \hat{X}_2/n_2$ are the sample proportions of defective components, where n_1 and n_2 are the sample sizes. In cell A1, type the title 'Hypothesis Testing about Population Proportion Difference D=(p1-p2) when D=.08' and centre across cells A1 to D1.

Data summary (given)
Sample size n_1 = 300, n_2 = 300; $\hat{X}_1$ = 33, $\hat{X}_2$ = 84; and significance level α = .01. In cells C3 and D3, type column headings 'Sample1' and 'Sample 2'. In cells B4 to B6, type the labels 'Sample size n =', 'Sample Xhat =' and 'Sample proportion phat='. To input the given information, in cells C4 and C5, type the corresponding values **300** and **33;** and, in cells D4 and D5, type the corresponding values **300** and **84.** In cell C6 and D6, type the formulas **=C5/C4** and **=D5/D4,** respectively, to obtain the $\hat{p}_1$ and $\hat{p}_2$ values.

Task 1 We want to test whether p_1 is at least .08 (eight percent) less than p_2. Therefore, the null and the alternative hypotheses are

$$H_0: \quad p_1 - p_2 = -.08$$
$$H_A: \quad p_1 - p_2 < -.08$$

and D = -.08 under H_0. As the alternative hypothesis contains the 'less than sign', i.e. the rejection region is on the left tail, this is a (left) one tail test.

In cells B8 and B9, type the labels 'Ho: (p1 - p2) =', 'HA: (p1 – p2) <' and, in cells C8 and C9, type the values, **-.08** and **-.08**.

Task 2 As D = -.08 ≠ 0, to test a hypothesis about the difference between two population proportions, we use the test statistic (see Case 2, Section 11.7)

$$Z = \frac{(\hat{p}_1 - \hat{p}_2) - (p_1 - p_2)}{\sqrt{\frac{\hat{p}_1\hat{q}_1}{n_1} + \frac{\hat{p}_2\hat{q}_2}{n_2}}} \sim \text{Normal}(0,1)$$

In cells B11 and C11, type the labels 'Test Statistic:' and 'Z = [(phat1 – phat2) – (p1-p2)]/SQRT[(phat1qhat1/n1) +(phat2qhat2/n2)]'.

Task 3 In cell B13, type the label 'Level of significance =' and, in cell C13, type the given value **0.01**.

Task 4 ***Calculating the critical value***
As this is a (left) one tail test, for a test with 1 percent level of significance,

$$\text{Critical value } z_{critical} = -z_{\alpha} = -z_{.01}$$

In cell B15, type the label 'Critical value z(.01) ='. In this example, since H_A: p_1 - p_2 < -0.08, the critical region is on the left tail of the z-distribution with area .01. To obtain this value, in cell C15, type **=NORMSINV(.01)** or **=NORMSINV(C13)**. Now the $z_{critical}$ value (–2.3263) will appear in the selected cell C15.

The decision rule

Therefore the decision rule for the test is

Method A: **Reject H_0 if $z_{calculated} < z_{critical} = -2.326$**

Method B: **Reject H_0 if *p*-value < α = .01**

In cells B16 to B18 type the labels 'Decision Rule: ', 'Method A: Reject Ho if Zcal <' and 'Method B: Reject Ho if p-value <'. In cells C17 and C18, type **=C15** and **=C13**.

Task 5 ***Calculating the data-based value of the test statistic ($z_{calculated}$)***

To calculate the data-based value of the test statistic ($z_{calculated}$) defined in Task 2, we first calculate the denominator, which is an estimator for the standard error of ($\hat{p}_1 - \hat{p}_2$) and then calculate the data-based value of the test statistic. In cells B20 to B21, type the labels 'SE of (phat1–phat2)=' and 'Test statistic Z-calculated ='. To obtain the standard error of ($\hat{p}_1 - \hat{p}_2$), in cell C20, type the formula **=SQRT((C6*(1-C6)/C4)+(D6*(1-D6)/D4))**.

To calculate the data-based value of the test statistic, in cell C21, type the formula **=(C6-D6-C8)/C20** or **=STANDARDIZE(C6-D6,C8,C20)**. This will give the $z_{calculated}$ value (= -2.848) of the test statistic.

Calculating the p-value

For a left tail z-test, the *p*-value = P(Z < $z_{calculated}$), where $z_{calculated}$ is the data-based value of the test statistic calculated above.

In cells B22 and B23, type the labels 'Cumulative probability =' and 'p-value='. Now we use the function **NORMSDIST** or **NORMDIST** to obtain the cumulative probability corresponding to $z_{calculated}$ = -2.848. The *p*-value for this (left) one tail test will be this cumulative probability. In cell C22, type **=NORMSDIST(-2.848)** or **=NORMSDIST(C21)** or **NORMDIST(C21,0,1,TRUE)** or **=NORMDIST(C6-D6,C8,C20,TRUE)**. This will give the required *p*-value for a (left) one tail test. In cell C23, type the cell reference **=C22**. This will give the *p*-value .002.

Task 6 ***Conclusion***

(a) Method A From Task 5, $z_{calculated} = -2.848$

From Task 4, $z_{critical} = -2.326$

Based on the decision rule developed in Task 4, as $z_{calculated} = -2.848 < z_{critical} = -2.326$, reject H_0: $p_1 - p_2 = -.08$ in favour of the alternative H_A: $p_1 - p_2 < -.08$. In cells B25 and C25, type the text, 'Based on Method A' and 'Since Zcal = -2.848 < Zcrit = -2.326, reject Ho'.

(b) Method B From Task 5, p-value = .002

Based on the decision rule given in Task 4, as the level of significance $\alpha = .01 > p$-value, reject H_0: $p_1 - p_2 = -.08$ in favour of the alternative H_A: $p_1 - p_2 < -.08$. In cells B26 and C26, type the text, 'Based on Method B' and 'Since p-value = .002 < alpha = .01, reject Ho'.

There is sufficient evidence to conclude that the proportion of defective components produced by innovation 1 is at least 8% smaller than the proportion of defective components produced by innovation 2. That is, the firm should adopt innovation 1.

Task 7 Save your workbook as **A:EX117.XLS** on your disk and print the file. The output from Tasks 1-6 is given in Figure 11.13.

Figure 11.13 Hypothesis testing about $p_1 - p_2 = D$ when $D \neq 0$

	A	B	C	D	E	F
1	Hypothesis Testing about Population Proportion Difference D = (p1-p2) when D=-.08					
2						
3	Data Summary:		Sample 1	Sample 2		
4		Sample size n =	300	300		
5		Sample Xhat =	33	84		
6		Sample proportion phat =	0.11	0.28		
7						
8	Step 1	Ho:(p1-p2) =	-0.08			
9		HA: (p1-p2)<	-0.08			
10						
11	Step 2	Test statistic:	Z=[(phat1-phat2) - (p1-p2)]/SQRT[(phat1qhat1/n1)+(phat2qhat2/n2)]			
12						
13	Step 3	Level of significance =	0.01			
14						
15	Step 4	Critical value z(.01) =	-2.326341928			
16		Decision rule:				
17		Method A: Reject Ho if Zcal <	-2.326341928			
18		Method B: Reject Ho if p-value <	0.01			
19						
20	Step 5	SE of (phat1-phat2) =	0.031596413			
21		Test statistic Z-calculated =	-2.848424571			
22		Cumulative probability =	0.002196877			
23		p-value =	0.002196877			
24						
25	Step 6	Based on Method A	Since Zcal =-2.848 < Zcrit = -2.326, reject Ho			
26		Based on Method B	Since p-val = .002 < alpha = .01, reject Ho			

11.10 Exercises to Chapter 11

Exercise 11.1

The following are summary exam marks of two Year 12 classes from two different schools (scored out of 200). Assume that the two populations are normal with known standard deviations $\sigma_1 = 10$ and $\sigma_2 = 15$.

Sample 1:	$n_1 = 15$	$\overline{X}_1 = 140$
Sample 2	$n_2 = 10$	$\overline{X}_2 = 150.$

Test the claim that the students from School 2 have performed better than those from School 1 (with $\alpha = .10$). That is, test the null hypothesis H_0: $(\mu_1 - \mu_2) = 0$ against the alternative H_A: $(\mu_1 - \mu_2) < 0$.

(The following exercises are based on exercises from Chapter 11 of Selvanathan et al., Australian Business Statistics, 1994)

Exercise 11.2

Kool Kat, a manufacturer of automobile air conditioners, is considering switching its supplier of condensers. Supplier *A*, the current producer of condensers for Kool Kat, prices its product 5% higher than supplier *B* does. Since Kool Kat wants to maintain its reputation for quality, however, Kool Kat wants to be sure that supplier *B*'s condensers last at least as long as supplier *A*'s do. The management of Kool Kat has decided to retain supplier *A* if there is sufficient statistical evidence that supplier *A*'s condensers last longer on the average than supplier *B*'s condensers. In an experiment, 10 midsize cars were equipped with air conditioners using type *A* condensers while another 10 midsize cars were equipped with type *B* condensers. The number of kilometers driven by each car before the condenser broke down was recorded, and the relevant statistics are listed in the accompanying table.

Type *A* Condensers	Type *B* Condensers
$\overline{X}_1 = 75{,}000$	$\overline{X}_2 = 70{,}000$
$s_1 = 6{,}000$	$s_2 = 5{,}000$

Assuming that the distance traveled is normally distributed, should Kool Kat retain supplier *A*? (Use $\alpha = .10$)

Exercise 11.3

Disposable batteries are expensive and they release dangerous chemicals when discarded. Consequently, many people are now using rechargeable batteries. However, some rechargeable batteries do not accept a full charge and as a result do not function as well as others. In an experiment to determine which batteries accept charges better, 100 D-cells made by firm *A* and another 100 D-cells made by firm *B* were randomly selected. Each battery was charged for 14 hours. The number of volts of power each was capable of producing was measured (D-cells are supposed to produce 1.25 volts); the results are summarized in the accompanying table. Do these results allow us to conclude at the 1% significance level that the mean power differs between the two brands of batteries?

Power Output (volts)	
Firm *A*	**Firm *B***
$\bar{X}_1 = 1.16$	$\bar{X}_2 = 1.21$
$s_1 = .08$	$s_2 = .10$

Exercise 11.4

A restaurant located in an office building decides to adopt a new strategy for attracting customers to the restaurant. Every week it advertises in the city newspaper. In the 10 weeks immediately prior to the advertising campaign, the average weekly gross was \$10 500, with a standard deviation of \$750. In the 8 weeks after the campaign began, the average weekly gross was \$12 000, with a standard deviation of \$1,000.

a. Assuming that the weekly grosses are normally distributed, can we conclude, with $\alpha =$ .10, that the advertising campaign was successful?
b. Assume that the net profit is 20% of the gross. If the ads cost \$100 per week, can we conclude, with $\alpha = .10$, that the ads are worthwhile?

Exercise 11.5

In a random sample of 500 television sets from a large production line, there were 80 defective sets. In a random sample of 200 television sets from a second production line, there were 10 defective sets. Do these data provide sufficient evidence to establish that the proportion of defective sets from the first line exceeds the proportion of defective sets from the second line by at least 3%? (Use $\alpha = .05$).

Exercise 11.6

A survey to study the usefulness of computers used a sample of 330 managers, professionals, and executives. This group included 132 computer users, 102 people who did not use computers, and 96 who said they would use them relatively soon. Asked if computers boost personal productivity, 50% of the non-users and intenders said *no*, as did 25% of the users. Can we conclude from these data, at the 1% level of significance, that users and non-users (including intenders) differ in their opinion of the usefulness of computers?

Chapter 12

Correlation and Regression Analysis

The statistical technique which is used to analyse the relationship among two or more variables whose data types are quantitative (or qualitative but transformed quantitative) is called *Regression Analysis*. One of the reasons for the importance of regression analysis is that it can be used to measure the strength of a relationship among variables and to forecast variables.

Aim: At the completion of the examples in this chapter, you should be able to use EXCEL to estimate the strength of the relationship between two variables and to estimate a linear relationship between two or more variables.

12.1 Analysing the relationship between two variables

The strength of a linear relationship between two variables can be measured by the coefficient of correlation (ρ). The correlation coefficient can take values in the range -1 to +1. If the two variables move in the same direction (that is, when X decreases, Y also decreases; or when X increases, Y also increases) then the coefficient of correlation is positive. If the two variables move in the opposite direction (i.e. when X decreases, Y increases; or when X increases, Y decreases) the coefficient of correlation is negative. If the movement of one variable does not affect the other, then the correlation between the two variables will be zero. In EXCEL, to calculate the coefficient of correlation, we use the statistical function **CORREL**.

To estimate the linear relationship between two variables X and Y in the form $Y = \beta_0 + \beta_1 X + \varepsilon$ (where Y is called the dependent variable and X is the independent variable and ε is a stochastic error term) we use the **least squares method** that minimizes the sum of squares errors (SSE). In this model, the dependent variable Y is influenced by only one independent variable X. Such a model is called a **simple regression model**.

In EXCEL, to estimate a regression model, we can use the Data Analysis Tool, **Regression** from the *Data Analysis...* option of the *Tools* menu. The summary output obtained from *Regression* contains the least squares estimation results as well as confidence intervals, values of the test statistics, *p*-values, etc. for the regression model parameters β_0 and β_1. Furthermore, the results can also be used to forecast the dependent variable Y.

The fit of the estimated regression model can be assessed using some of the summary measures in the *Regression* output. The two important summary measures used to assess the fit of the regression model are the coefficient of determination R^2 and the standard error of estimate s_e. The coefficient of determination R^2 can take any value between 0 and 1. A value of R^2 close to 1 indicates an excellent fit and a value close to 0 indicates a poor fit.

There are a number of assumptions the error terms should satisfy in order to have desirable qualities such as unbiasedness and efficiency for the estimators β_0 and β_1 (usually known as BLUE). Five such important assumptions on the error terms are, namely, (1) errors have zero mean; (2) errors are independent of the independent variables; (3) errors have constant variance (or errors are homoscedastic); (4) errors are serially uncorrelated; and (5) errors are normally distributed. Usually, assumption 3 could be violated when we use cross-sectional data for estimation and assumption 4 could be violated when we deal with time-series data. A well-known and simple test for serial correlation is the Durbin-Watson test.

12.2 Example 12.1: Simple linear regression

The accompanying table exhibits, for 8 delicatessens, the annual profit per dollar of sales (measured in cents), Y, and the number of employees per store, X.

X	3	6	4	5	2	5	4	1
Y	19	30	20	25	16	26	22	14

Task 1 Enter the data in cells A1 to B9 in two columns (X in column A and Y in column B).

Task 2 Draw a scatter plot of profit per dollar of sales (Y) against the number of employees per store (X).

Task 3 Measure the strength of the linear relationship between profit per dollar of sales (Y) and the number of employees per store (X).

Task 4 Estimate a linear relationship between profit per dollar of sales and the number of employees per store in the form

$$Y = \beta_0 + \beta_1 X + \varepsilon.$$

Obtain the residuals and fitted values. Obtain the plot of Y and predicted value of Y against X on the same graph.

Task 5 Draw the estimated regression line obtained in Task 4 on the same plot from Task 2.

Task 6 What is the value of the coefficient of determination R^2? What is the standard error of estimate s_e? Does it appear that a simple linear regression model is reasonable?

Task 7 Calculate the 95% confidence interval for the slope coefficient β_1.

Task 8 Is the relationship significant at the 5 percent level of significance?

Task 9 Plot the actual values against fitted values and comment on the predictability of your estimated model.

Task 10 Forecast the profit per dollar sales of a delicatessen which employs 7 people.

Task 11 Plot the residual values against X and comment on the issue of heteroscedasticity.

Task 12 Save your workbook as 'A:EX121.XLS' on your disk and print the file.

Solution to Example 12.1

Task 1 Enter column headings '**X**' and '**Y**' in cells A1 and B1, respectively, and the values for the two variables in cells A2 to A9 and B2 to B9. [Note: To plot the variable Y against variable X, the variable on the (horizontal) X-axis should be typed in column A and the variable on the (vertical) Y-axis should be typed in column B.]

Task 2 ***Drawing a scatter plot***

To draw a scatter plot in EXCEL, first highlight the data in cells A1 to B9. Click the **ChartWizard** on the *Standard toolbar*. Then click at the point where you want the graph on the worksheet. As you already know, there are 4 steps in **ChartWizard**. In Step 1, select the *Chart type* **XY(scatter)** and *Chart sub-type* **Scatter** (first selection in row 1) and then click **Next>** to go to Step 2. In the Step 2 dialogue box, EXCEL will show the *Data range* as **$A$1:$B$9.** [If you have not already highlighted the data in cells A1 to B9, the space for *Data range* will be blank. If this is the case, simply type **A1:B9** in the blank space.] For *Series in*: click on **Columns**. To go to Step 3, click **Next>**. In Step 3, give (i) a *chart title* (for example, type 'Annual Profit per Dollar of Sales'); (ii) *value [x] axis* (for example, 'Number of Employees') and *value [y] axis* (for example, 'Profit'). On the left, EXCEL will display a sample chart. If you are satisfied with the data series information in the dialogue box, click **Next>** to go to Step 4, otherwise change the information and then click **Next>**. In Step 4, click on the circle for *As Object in:* and finally, click **Finish**. The resulting plot can be seen in Figure 12.1.

Figure 12.1 Graph of annual profit against the number of employees

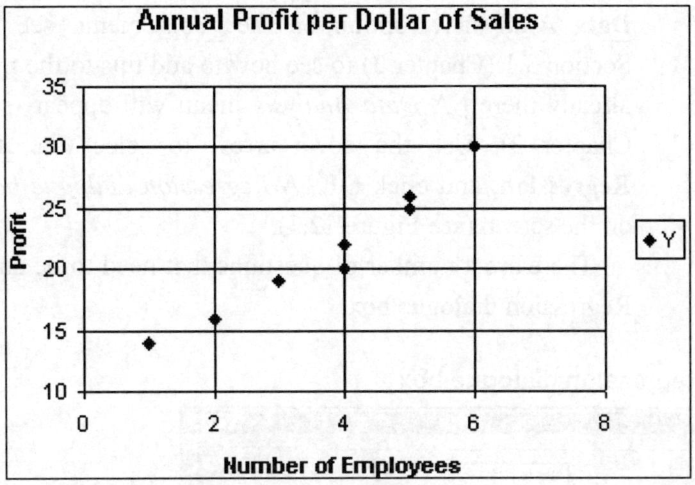

Task 3 ***Calculating the coefficient of correlation***

To measure the strength of the linear relationship between two variables X and Y, we can use the correlation coefficient between X and Y. To obtain the coefficient of correlation between X and Y, first, in cell A11, type the label 'Correlation(X,Y)=' and click on cell B11 to store the correlation value. Now click the icon **Paste Function** on the *Standard toolbar* and select *Function Category* **Statistical** and *Function Name* **CORREL** and then click **OK**. The dialogue box appearing on the screen (see Figure 12.2) will ask for two input values, namely, *Array 1* (location of values of X) and *Array 2* (location of values of Y). Type **A2:A9** in the space for *array 1*, **B2:B9** in the space for *array 2* and click on **Finish**. The correlation value 0.976782 will be stored in the selected cell B11. Alternatively, in cell B11, type the formula **=CORREL(A2:A9,B2:B9)**.

Figure 12.2 CORREL Function dialogue box

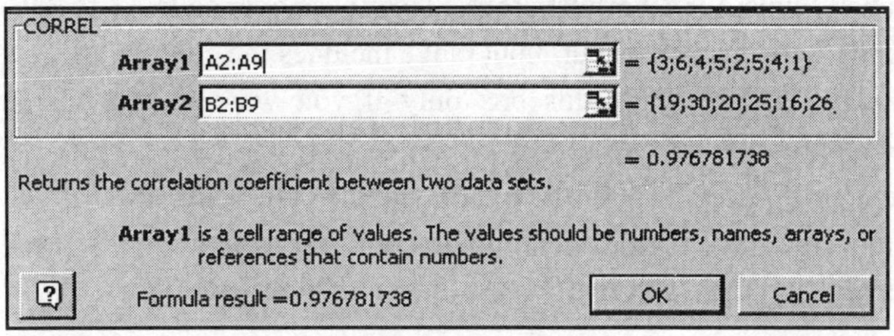

Task 4 ***Estimating a linear regression line***

There are a number of ways a linear regression model can be estimated using EXCEL. For this task, we illustrate the three main methods.

Method 1 To estimate the linear relationship between two variables X and Y, we can use the **Tools** option on the *Menubar*. Click the *Tools* option on the *Menubar* and select **Data Analysis...** [If you cannot find the **Data Analysis...** option on the *Tools* menu, see footnote 1 in Section 3.1 (Chapter 3) to see how to add this to the menu if it is not already there.] A *Data Analysis* menu will appear (see Figure 3.1, Chapter 3). Use the down arrow to select the *Analysis Tool*, **Regression**, and click **OK.** A *Regression dialogue box* will appear on the screen (see Figure 12.3).

There are a number of questions that need to be answered on the Regression dialogue box.

Figure 12.3 Regression dialogue box

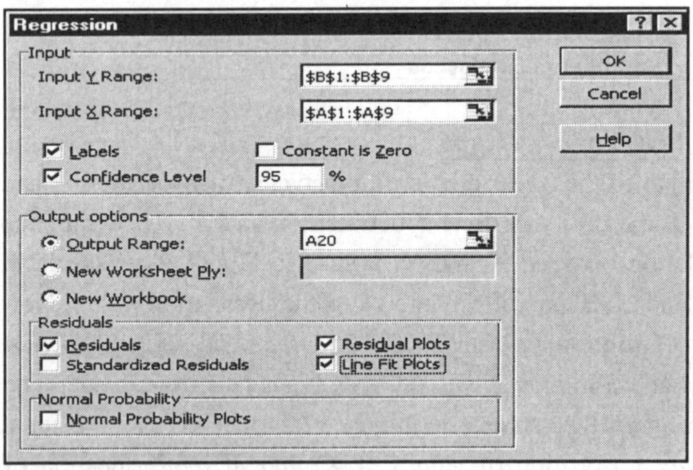

Input

Input Y Range: Enter the input range for Y as **B1:B9**. You can manually type **B1:B9** or just browse the mouse over cells **B1 to B9**.

Input X Range: Enter the input range for X as **A1:A9**. You can manually type **A1:A9** or just browse the mouse over cells **A1 to A9**.

Labels: Tick this box as our input range includes the column title.

Constant is Zero: Tick this box only if you want to restrict the intercept term to be zero.

Confidence Level: Tick this box and type the confidence level if you require the confidence intervals for the regression coefficients, intercept and the slope coefficients (as required in Task 7).

Output Options

Output Range: Click on the little circle against **Output Range** and type the output range[1] (the area where you want the output to be displayed), **A20**, say. Starting point for the output range is sufficient.

[1] To avoid overwriting the raw data you have in cells A1 to B9 and the output from Tasks 2-3 with the regression output, the Output Range should not overlap the Input Range and the output from Tasks 2-3.

New Worksheet Ply: The output can be saved on a separate workbook if you select the **New Workbook** option.

New Workbook: The output can be saved on a separate workbook if you select the **New Workbook** option.

Residuals

Residuals: Tick this box since we require the residuals. [These residuals are required for the residual plot to check serial correlation.]

Residual Plots: Tick this box since we require the residual plots against X-variable. [This plot is required for Task 11 to check heteroscedasticity.]

Standardised Residuals: Tick this box only if you require the standardised residuals.

Line Fit Plots: Tick this box as we require the scatter plots of the actual and predicted values of the Y-values against the X-values.

Normal Probability Plot: Tick this box only if you require a normal probability plot of the residuals.

Click **OK**. The summary regression output is presented in Figure 12.4.

Method 2 To estimate the linear relationship between two variables X and Y and to obtain some of the summary statistics, first **highlight a 5×2 cell area** (for example, cells **D20 to E24**) on the worksheet. Then click on the icon **Paste Function** on the *Standard toolbar*. Then select the *Function Category* **Statistical** and the *Function Name* **LINEST**. [This function will return estimates of the slope (b_1) and intercept term (b_0) of the regression line, standard errors of the coefficients (s_{b1}, s_{b0}), coefficient of determination (R^2), standard error of estimate (s_e), F statistic, degrees of freedom, SSR and SSE]. Then click **OK** . The dialogue box appearing on the screen will ask for 4 inputs (see Figure 12.5).

Known y's:	Range for the y values;
Known x's:	Range for the x values;
Constant:	A logical value **TRUE** if there is a constant term, **FALSE** otherwise
Stats:	A logical value **TRUE** if additional summary statistics are required, **FALSE** otherwise.

Now, type the four input values **B2:B9**, **A2:A9**, **TRUE**, **TRUE** in the allocated spaces on the dialogue box and then click on **OK**. Now you will see the slope coefficient 3.128205 to appear on cell D20 and the other cells still empty. Then go to the *Formula bar* (located just above the column headings A, B, C etc.) and click the

Figure 12.4 Summary regression output

	A	B
20	SUMMARY OUTPUT	
21		
22	*Regression Statistics*	
23	Multiple R	0.976781738
24	R Square	0.954102564
25	Adjusted R Square	0.946452991
26	Standard Error	1.236897138
27	Observations	8

	A	B	C	D	E	F	G
29	ANOVA						
30		*df*	*SS*	*MS*	*F*	*Significance F*	
31	Regression	1	190.820513	190.821	124.726	3.07493E-05	
32	Residual	6	9.17948718	1.52991			
33	Total	7	200				
34							
35		*Coefficients*	*Standard Error*	*t Stat*	*P-value*	*Lower 95%*	*Upper 95%*
36	Intercept	9.7692308	1.13777988	8.58622	0.00014	6.985181652	12.55328
37	X	3.1282051	0.28010204	11.1681	3.1E-05	2.442819615	3.8135906

	A	B	C
41	RESIDUAL OUTPUT		
42			
43	*Observation*	*Predicted Y*	*Residuals*
44	1	19.1538462	-0.153846154
45	2	28.5384615	1.461538462
46	3	22.2820513	-2.282051282
47	4	25.4102564	-0.41025641
48	5	16.025641	-0.025641026
49	6	25.4102564	0.58974359
50	7	22.2820513	-0.282051282
51	8	12.8974359	1.102564103

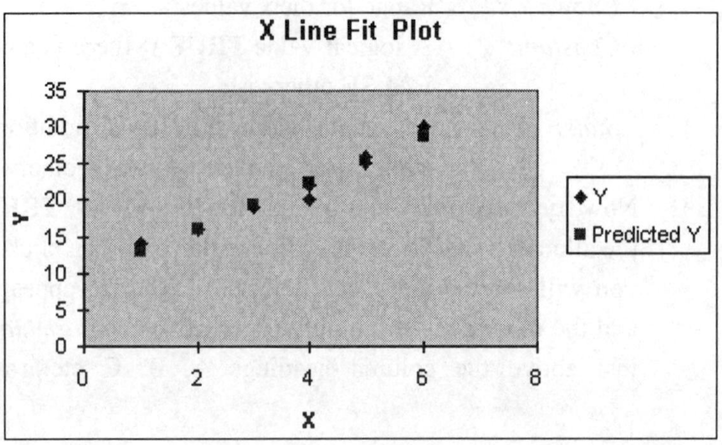

Figure 12.5 LINEST Function dialogue box

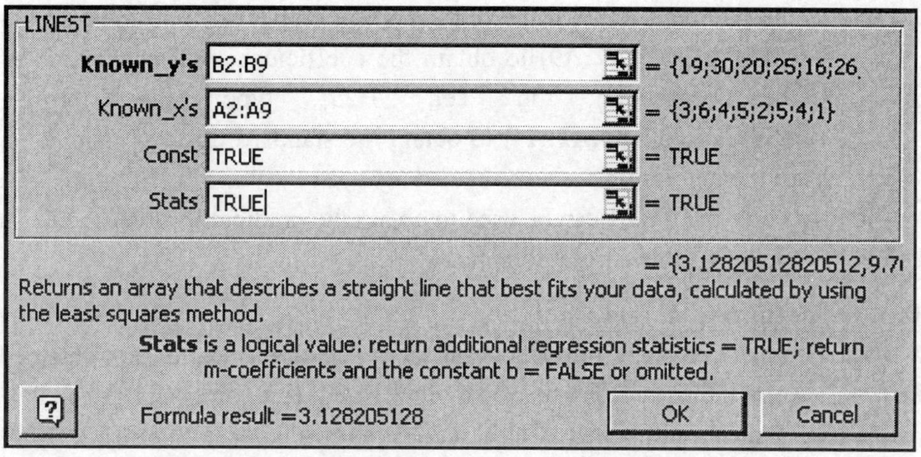

mouse at the end of the formula in the *Formula bar* after '**TRUE)**' and press **CTRL + SHIFT + ENTER** keys simultaneously. You will notice the 5×2 block of cells D20 to E24 filled with numbers as in the table below.

Alternatively, in cell D20, you can type **=LINEST(B2:B9,A2:A9,TRUE,TRUE)**. Now highlight a 5×2 cells area starting with cell D20 (i.e. cells D21 to E24) to store the results on the worksheet. Then click the mouse at the end of the formula in the *Formula bar* after **TRUE)** and press **CTRL + SHIFT + ENTER** keys simultaneously.

The results will appear in the following format:

Cell	D		E	
20	Slope Coefficient (b1)	**3.128**	Intercept Term (bo)	**9.769**
21	Standard error of b1(Sb1)	**.280**	Standard error of bo (Sbo)	**1.14**
22	R^2	**.954**	S_e	**1.237**
23	F statistics	**124.726**	Degrees of freedom	**6**
24	SSR	**190.821**	SSE	**9.179**

Method 3 The slope coefficient, intercept, the coefficient of determination R^2 and standard error of estimate for a simple regression model can also be individually obtained by typing the formulas in the form **FUNCTION NAME(*known y's; known x's*)** in the appropriate cells.

In cells G20 to G23, type the labels 'Slope=', 'Intercept =', 'R-Square=' and 'Se='. In cell H20, type the formula **=SLOPE(B2:B9,A2:A9)** to obtain the slope of the regression; in cell H21, type the formula =**INTERCEPT(B2:B9,A2:A9)** to obtain the intercept term of the regression; in cell H22, type the formula =**RSQ(B2:B9,A2:A9)** to obtain the coefficient of determination of the regression; in cell H23, type the formula =**STEYX(B2:B9,A2:A9)** to obtain the standard error of estimate s_e of the regression.

Note that only Method 1 can be used to obtain the predicted values, residuals and plots required in this task, automatically.

Task 5 ***Drawing the estimated regression line***

We can automatically superimpose a regression line on the scatter plot obtained in Task 2. Point the mouse on one of the highlighted data points and **right-click** the mouse. A menu will appear. Select **Add Trendline...** (see Figure 12.6)]. A *Trendline dialogue box* will appear (see Figure 12.7).

Figure 12.6 Trendline menu

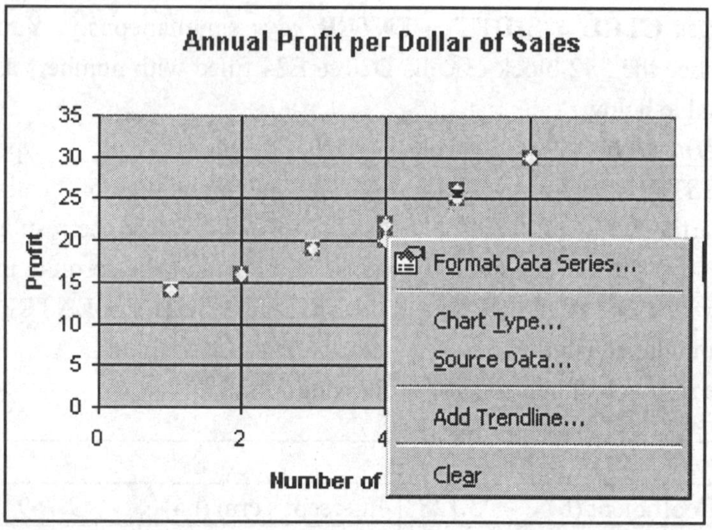

Figure 12.7 Trendline Type dialogue box

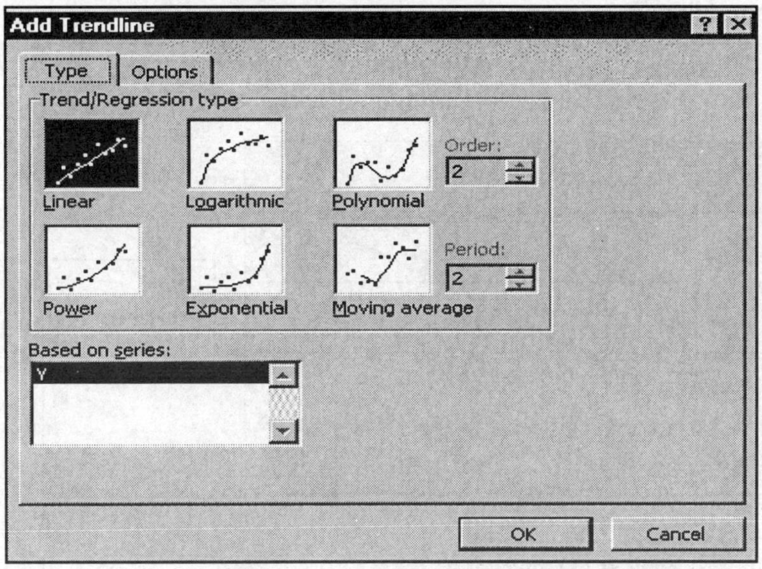

On the *Trendline dialogue box*, click on **Type** and select *Trend/Regression type* **Linear**. Click **Options** on the same dialogue box and another *Trendline dialogue box* will appear on the screen (see Figure 12.8). Now click the little square boxes next to *Display equation on Chart* and *Display R-Squared Value on Chart*. Then click **OK**. The output is given in Figure 12.9.

Figure 12.8 Trendline Options dialogue box

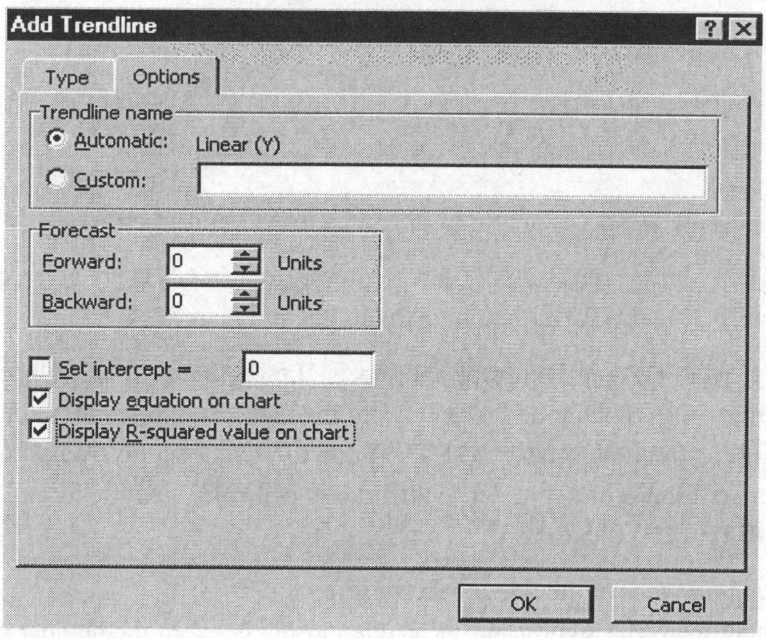

Figure 12.9 Graph of annual profit against the number of employees and the linear trend line

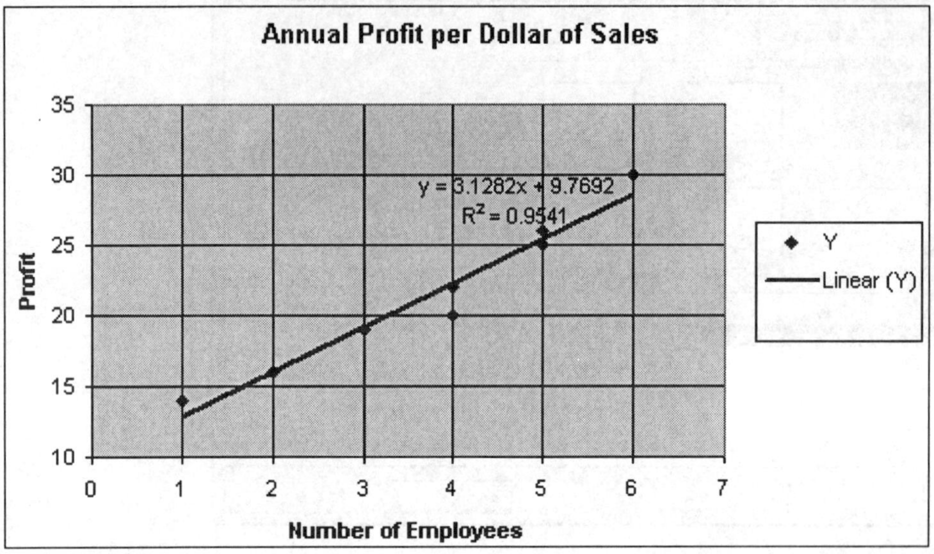

Task 6 From the Regression output in Figure 12.4, we have, the coefficient of determination $R^2 = .95$ and the standard error of estimate $s_e = 1.23$. Yes. From the plot in Figure 12.10 and the high value of R^2 as well as the low s_e value (relative to the Y values), we can conclude that a straight line model is reasonable.

Task 7 ***Constructing confidence interval for the slope coefficient***

95% confidence interval for the slope coefficient is given by

$$b_1 \pm t_{\alpha/2,n-2} * s_{b_1}$$

$$= b_1 \pm t_{.025,6} * s_{b_1}$$

where b_1 is the least squares estimator of β_1 and s_{b1} is the standard error of b_1.

Method 1 These upper limit (=3.814) and the lower limit (=2.443) are given in cells G37 and F37 of the summary output, in Figure 12.4.

Method 2 In cells F42 and F43, type the labels 'Lower limit =' and 'Upper limit ='. To obtain the lower limit of the confidence interval, in cell G42, type the formula **=B37-TINV(.05,6)*C37**. Then to obtain the upper limit of the confidence interval, in cell G43, type the formula, **=B37+TINV(.05,6)*C37**.

Task 8 ***Testing the significance of the linear relationship***

To test whether there is a significant linear relationship between the number of employees (X) and profit (Y), the null and the alternative hypotheses are given by

H_o: $\beta_1 = 0$
H_A: $\beta_1 \neq 0$.

Method 1 The value of the test statistic, $t_{calculated} = b_1 / s_{b_1}$ (=11.16809), is given in cell D37 of the summary output, in Figure 12.4. Now we obtain the $t_{critical}$ value using the function TINV. In cell F40, type the label 't critical ='. To obtain the $t_{critical}$ value, in cell G40, type the formula **=TINV(.05,6)**. This will give $t_{critical}$ = 2.447. As $t_{calculated}$ = 11.16809 > $t_{critical}$ = 2.447, we reject H_o: $\beta_1 = 0$ in favour of the alternative H_A: $\beta_1 \neq 0$, and conclude that there exists a significant linear relationship between X and Y.

Method 2 The *p*-value (= 3.1E-05) of the test is given in cell E37 of the summary output in Figure 12.4. Comparing this with the level of significance α = .05, we have *p*-value = .000031 < .05 = α. Therefore, we reject H_o: $\beta_1 = 0$ in favour of the alternative H_A: $\beta_1 \neq 0$, and conclude that there exists a significant linear relationship between X and Y.

Therefore, based on the sample information, we conclude that there exists a significant linear relationship between the number of employees and profit.

Task 9 ***Graphing the actual and predicted values***

In cell C1, type the label, 'Yhat'. Copy the predicted values in cells B44 to B51 (from the summary output in Figure 12.4) onto cells C2 to C9. To draw the plot of the actual values (in cells B1 to B9) against the predicted values (in cells C1 to C9), we highlight cells B1 to C9 and use the **ChartWizard** as described in Task 2. To see the level of predictability of the model, as in Task 5, we superimpose an estimated linear trend line with zero intercept to show the relationship between Y and Yhat. If the slope of this line is close to one, then it can be considered as the line through the origin at an angle 45°.

To draw the linear trend line, we follow the same procedure as in Task 5 except in the *Trendline dialogue box* for **Options** (see Figure 12.8). For this task, tick the little square boxes appearing next to *Set Intercept= 0, Display equation on Chart* and *Display R-Squared Value on Chart*. The output (with reformatted X-axis) is given in Figure 12.10.

As can be seen, the slope of the estimated linear relationship between Y and Yhat is .9976, which is very close to 1, indicating that the points on the plot lie very close to a 45° line through the origin. This shows that the predictability of the estimated model is excellent.

Task 10 ***Forecasting the dependent variable***

Method 1 In cell A13, type the label 'Forecast Y(X=7) ='. To obtain the forecast for profit per dollar sales of a delicatessen which employs 7 people, type the formula **=B36+B37*7** in cell B13.

Figure 12.10 Graph of predicted against actual profit

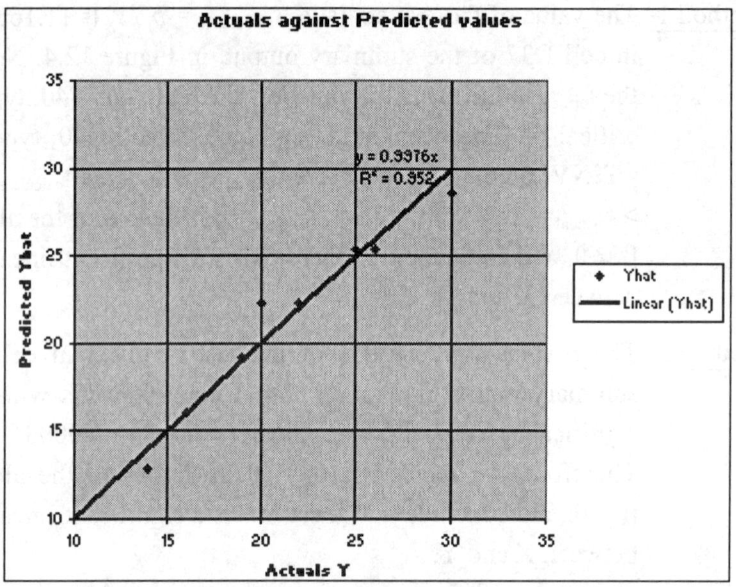

Method 2 Click on the **Paste Function** icon on the *Standard toolbar*. In Step 1, select *Function Category* **Statistical** and *Function Name* **FORECAST** and then click **OK**. Now you will see a forecast dialogue box (see Figure 12.11) appearing on the screen asking for 3 inputs: *x* (the data point for which a predicted value is required); *known y's* (range of Y); *known x's* (range of X). Type the entries **7**; **B2:B9** and **A2:A9** as the three required inputs and click **OK**. *Alternatively*, we can also use the statistical function FORECAST in the form FORECAST(new x's, known y's, known x's). In this case, type the formula **=FORECAST(7,B2:B9,A2:A9)** in cell B13.

Figure 12.11 Forecast dialogue box

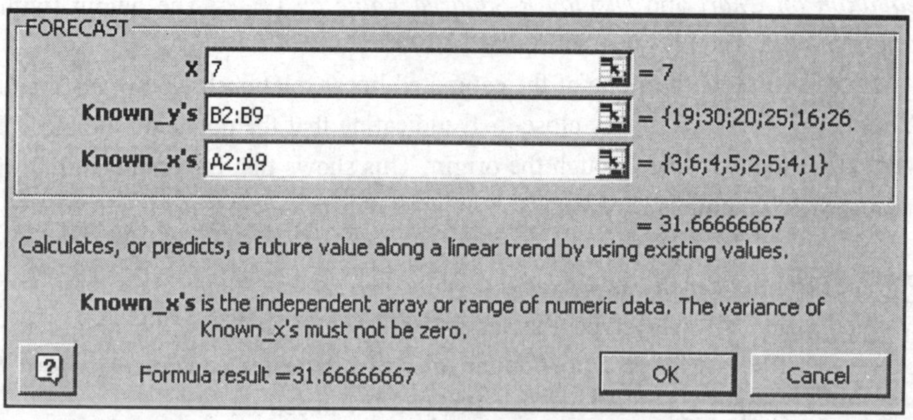

Task 11 ***Checking for heteroscedasticity***

Figure 12.12 presents the plot of the residuals against the variable X (this plot is given in the output from the Analysis Tool *Regression* in Task 4 when *estimating a linear regression line using Method 1*).

Figure 12.12 Graph of residual against X

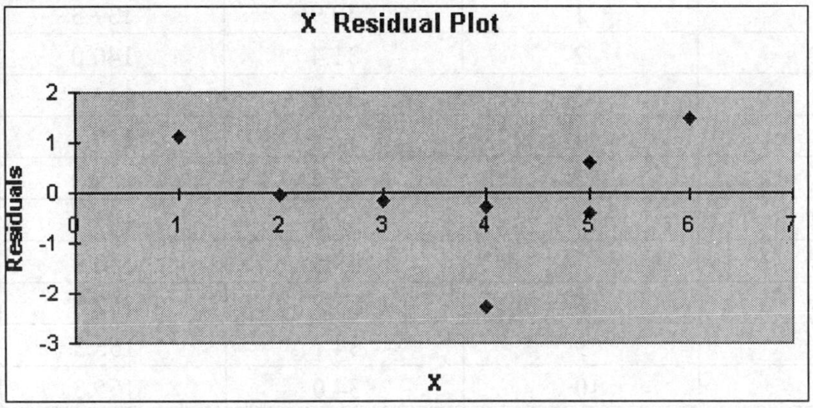

As can be seen from Figure 12.12, the variance (spread of the plot) appear to be a constant as most of the points fall within a rectangular band. Hence, there is no evidence for strong heteroscedasticity. That is, the errors appear to be homoscedastic.

Task 12 Save your workbook as **A:EX121.XLS** on your disk and print the file.

12.3 Analysing the relationship between two or more variables

In Section 12.1, the dependent variable Y is influenced by only one independent variable X, and the relationship is called a simple regression model. When the dependent variable Y is influenced by more than one independent variables X_1, X_2,..., X_k, then the relationship can be written in a linear form as

$$Y = \beta_0 + \beta_1 X_1 + \beta_2 X_2 + \beta_3 X_3 + \dots + \beta_k X_k + \varepsilon.$$

Such a model is called a **multiple regression model**. In EXCEL, the estimation of the multiple regression model can be done in the same way as a simple regression model using the *Data Analysis tool*, **Regression**.

12.4 Example 12.2: Multiple linear regression

The data below show the sales in millions of dollars (X) of a major retail store in Sydney and the sales of all its chain stores around the country (Y) over a 20 quarterly period.

Quarter	Period (t)	X	Y
1	1	30.9	137.3
2	2	31.4	140.0
3	3	31.9	142.7
4	4	31.5	139.4
1	5	32.4	145.0
2	6	32.8	147.1
3	7	33.5	150.3
4	8	33.7	152.8
1	9	34.1	155.5
2	10	34.0	155.3
3	11	34.5	158.3
4	12	34.8	156.4
1	13	35.0	160.2
2	14	35.6	163.1
3	15	37.4	167.3
4	16	37.0	170.7
1	17	37.5	174.2
2	18	37.8	175.6
3	19	38.2	178.7
4	20	38.8	181.9

Task 1 Enter the data in cells A1 to C21 in three columns (*t* in column A, X in column B and Y in column C) with column headings 't', 'X' and 'Y' in cells A1, B1 and C1.

Task 2 Draw a scatter plot of Y against X. Comment on the plot.

Task 3 Measure the strength of the linear relationship between Y and X.

Task 4 Estimate a multiple linear relationship between Y, X and *t* in the form

$$Y = \beta_0 + \beta_1 t + \beta_2 X + \varepsilon.$$

Obtain the residuals, a residual plot and a line fit plot.

Task 5 Construct a 95% confidence interval estimate for the coefficient β_1.

Task 6 Are the estimated coefficients individually significant at the 5 percent level of significance?

Task 7 What is the value of the coefficient of determination R^2? What is the standard error of estimate s_e? Does it appear that the above model is reasonable? Use a test to assess the overall utility of the model.

Task 8 From the plot of the residuals, explain whether you find evidence of positive autocorrelation. Conduct a formal test for positive autocorrelation using Durbin-Watson test at significance level .01. Comment on whether the residual plot results are in accordance with the DW test results?

Task 9 Do you suspect heteroscedasticity of errors? Use the plot of the residuals against the X to confirm your observation.

Task 10 Plot the actual values against the fitted values and comment on the predictability of your estimated model.

Task 11 Forecast the total sales of all chain stores for the next four quarters if the sales of the major retail store in Sydney during these four quarters are expected to be 38.5, 38.8, 39.3 and 38.8 millions of dollars, respectively.

Task 12 Set up seasonal dummies for the 4 quarters with the base as the 4th quarter and estimate a model of the form

$$Y = \beta_0 + \beta_1 t + \beta_2 X + \beta_3 D_1 + \beta_4 D_2 + \beta_5 D_3 + \varepsilon.$$

Are the seasonal dummies significant?

Task 13 Save your EXCEL workbook as 'A:EX122.XLS' on your disk and print the file.

Solution to Example 12.2

Task 1 Enter column headings 't', 'X' and 'Y' in cells A1, B1 and C1, respectively, and the data for the three variables in cells A2 to A21, B2 to B21 and C2 to C21.

Task 2 ***Drawing a scatter plot for Y against X***

To draw a scatter plot for Y against X in EXCEL, follow the same procedure as in Task 2 of Example 12.1 and select the data in cells B1 to C21. The resulting plot is shown in Figure 12.13. As can be seen from the plot, there appears a strong linear relationship between Y and X.

Figure 12.13 Graph of Y against X

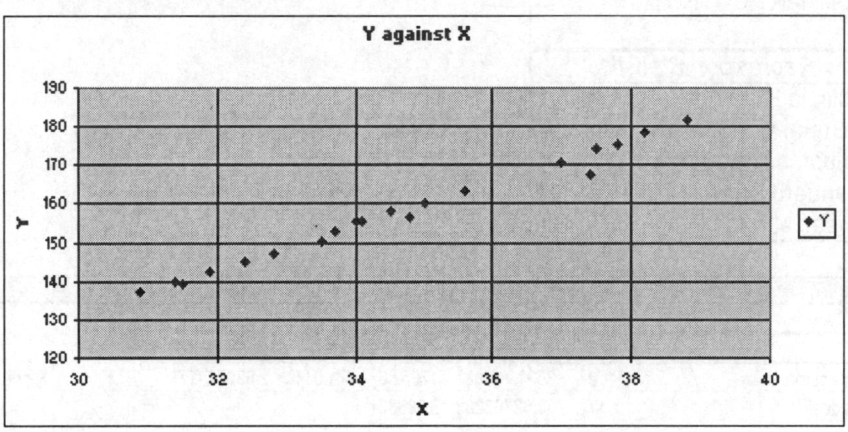

Task 3 ***Calculating the coefficient of correlation between Y against X***

To measure the strength of the linear relationship between two variables X and Y, we can use the correlation coefficient between X and Y. To obtain the correlation coefficient between X and Y, in cell A30, type 'Correlation(X,Y) =' and click on cell B30 to store the correlation value. In cell B30, type the formula **=CORREL(B2:B21,C2:C21)**.

Task 4 ***Estimating a Linear Regression Model***

To estimate the linear relationship between Y (Dependent variable) and a number of independent variables X_1, X_2 , …., we can use the Data Analysis tool *Regression*. Click the *Tools* option on the *Menubar* and select **Data Analysis...** Use the down arrow to select the *Analysis Tool*, **Regression**, and click on **OK.** The main difference between using the *Regression* tool between a simple regression model and a multiple regression model (such as this example) is that for the multiple regression model estimation, we should set up the data for all independent variables in adjacent columns as a block of columns.

On the *Regression dialogue box* under **Input,** enter the *Input Y Range* as **C1:C21**. Enter the *Input X Range* for X as **A1:B21**. Tick the box for *Label*: as our *Input Range* includes the column titles. Tick the box for *Confidence Level*: and specify the confidence level as 95% as we require the confidence intervals for the regression coefficients in Task 5. Under **Output Options,** click on the little circle against *Output Range* and type the output range (the area where you want the output to be displayed), **A35**, say. Starting point for the output range is sufficient. Under **Residuals,** click the box for *Residuals* as we require the residuals and click the box for *Residual Plots* as we require the residual plots (for Tasks 8 and 9). Click **OK.** The summary regression output is presented in Figures 12.14.

Figure 12.14 Summary multiple regression output

	A	B
35	SUMMARY OUTPUT	
36		
37	*Regression Statistics*	
38	Multiple R	0.99453
39	R Square	0.9891
40	Adjusted R Square	0.98781
41	Standard Error	1.51322
42	Observations	20

	A	B	C	D	E	F	G	H	I
44	ANOVA								
45		*df*	*SS*	*MS*	*F*	*Significance F*			
46	Regression	2	3531.110967	1765.56	771.044	2.0865E-17		F crit =	3.5915
47	Residual	17	38.92703268	2.28983					
48	Total	19	3570.038						

	A	B	C	D	E	F	G
49							
50		*Coefficients*	*Standard Error*	*t Stat*	*P-value*	*Lower 95%*	*Upper 95%*
51	Intercept	18.431513	28.0722637	0.6566	0.5202	-40.79587	77.658895
52	t	0.727329	0.38593174	1.8846	0.0767	-0.086917	1.5415749
53	x	3.796811	0.92576875	4.1013	0.0007	1.8436069	5.750015

	A	B	C	D
56	RESIDUAL OUTPUT			
57				
58	*Observation*	*Predicted y*	*Residuals*	
59	1	136.4803	0.8196987	
60	2	139.106	0.8939642	0.819699
61	3	141.7318	0.9682297	0.893964
62	4	140.9404	-1.540375	0.96823
63	5	145.0848	-0.0848339	-1.54037
64	6	147.3309	-0.2308873	-0.08483
65	7	150.716	-0.4159839	-0.23089
66	8	152.2027	0.5973248	-0.41598
67	9	154.4487	1.0512714	0.597325
68	10	154.7964	0.5036235	1.051271
69	11	157.4221	0.877889	0.503624
70	12	159.2885	-2.8884833	0.877889
71	13	160.7752	-0.5751745	-2.88848
72	14	163.7806	-0.6805901	-0.57517
73	15	171.3422	-4.0421788	-0.68059
74	16	170.5508	0.1492165	-4.04218
75	17	173.1765	1.023482	0.149217
76	18	175.0429	0.5571097	1.023482
77	19	177.2889	1.4110563	0.55711
78	20	180.2944	1.6056407	1.411056

Task 5 ***Constructing confidence intervals***

As can be seen from cells F52 and G52 of the output in Figure 12.14, the lower limit and upper limit of the 95% confidence of the coefficient β_1 are 1.844 and 5.750, respectively.

Task 6 ***Testing the significance of the linear relationship***

To test the significance of the individual coefficient estimates (i.e. to test the significance of the relationship between the dependent variable Y and the individual independent variables), we test

H_o: $\beta_i = 0$

H_A: $\beta_i \neq 0$, for i = 0,1,2.

Method 1 The value of the test statistic, $t_{calculated}=b_i / s_{b_i}$ for i=0,1,2, are given in cells D51 to D53 of the output in Figure 12.14. Now we obtain the $t_{critical}$ value using the statistical function TINV. In cell A31, type 't critical ='. To obtain the $t_{critical}$ value, in cell B31, type the formula **=TINV(.05,17)**. This will give $t_{critical} = 2.11$.

Testing for the significance of β_o: As $t_{calculated} = .66 < t_{critical} = 2.11$, do not reject H_0: $\beta_o = 0$ and conclude that the intercept term is insignificant.

Testing for the significance of β_1: As $t_{calculated} = 1.88 < t_{critical} = 2.11$, we do not reject H_0: $\beta_1 = 0$ and conclude that the trend term is insignificant.

Testing for the significance of β_2: As $t_{calculated} = 4.10 > t_{critical} = 2.11$, reject H_0: $\beta_2 = 0$ in favour of the alternative H_A: $\beta_2 \neq 0$ and conclude that there exists a significant linear relationship between X and Y.

Method 2 From cells E51 to E53 of the output, we see that the *p*-values of the three tests H_o: $\beta_i = 0$ against H_A: $\beta_i \neq 0$, i=0,1,2, are given as .52, .077 and .0007, respectively.

Testing for the significance of β_0: As *p*-value = .52 > level of significance $\alpha = .05$, we do not reject H_0: $\beta_o = 0$ and conclude that the intercept term is insignificant.
Testing for the significance of β_1: As *p*-value = .08 < level of significance $\alpha = .05$, we do not reject H_0: $\beta_1 = 0$ and conclude that the trend term is insignificant.
Testing for the significance of β_2: As *p*-value = .0007 < level of significance $\alpha = .05$, we reject H_0: $\beta_2 = 0$ in favour of the alternative H_A: $\beta_2 \neq 0$ and conclude that there exists a significant linear relationship between X and Y.

Task 7 ***Assessing the overall utility of the model***

From Task 4, we have the coefficient of determination $R^2 = .99$ (in cell B38), and the standard error of estimate $s_e = 1.51$ (in cell B41). Yes, high R^2 and low standard error of estimate s_e (relative to the sample mean of Y) suggest an excellent fit and that the estimated linear regression line is reasonable.

To assess the overall utility of the model, we test

H_o: $\beta_1 = 0$ and $\beta_2 = 0$
H_A: At least one $\beta_i \neq 0$, i=1,2.

Method 1 From the summary output (cell E46), $F_{calculated} = 771.04$. In cell H46, type 'F crit =' and, in cell I46, type the formula **=FINV(.05,2,17)**. This will give $F_{critical} = 3.59$. Since $F_{calculated} = 771.04 > F_{critical} = 3.59$, we reject H_o: $\beta_1 = 0$ and $\beta_2 = 0$ in favour of the alternative, and conclude that the model is useful.

Method 2 From the summary output (cell F46), the *p*-value to test H_o: $\beta_1 = 0$ and $\beta_2 = 0$ is shown as 2.09E-17 < $\alpha = .05$. Therefore, we reject H_o and conclude that the model is useful.

Task 8 ***Calculating the DW statistic and testing for zero serial correlation***

Figure 12.15 presents the residual plot.

Clearly, from the residual plot (see *t residual plot*), one would not suspect autocorrelated errors as the errors are scattered at random without showing any

Figure 12.15 Graph of residual against t

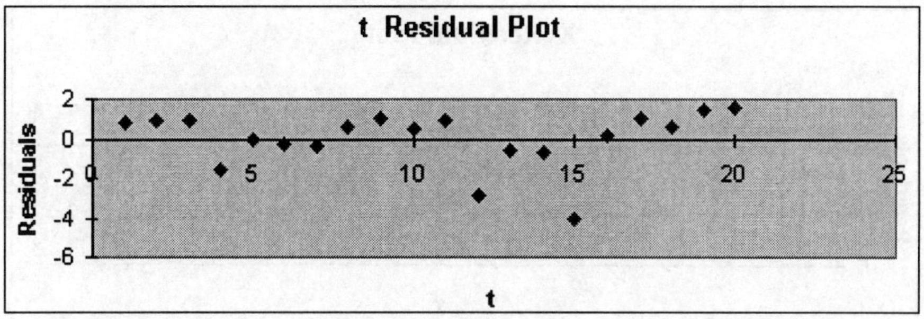

specific pattern. To formally test for serially correlated errors, we can use the Durbin-Watson (DW) test. We use EXCEL to calculate the approximate Durban-Watson statistic as

$$DW \cong 2(1-\rho),$$

where ρ is the slope of the regression

$$e_t = \alpha + \rho\, e_{t-1} + u_t$$

and e_t (t=1,2,3,…, 20) are residuals from the EXCEL summary output given in cells C59 to C78. To obtain the values of e_{t-1} (the lagged value of e_t), we copy entries in cells C59 to C78 onto cells D60 to D79. We then obtain ρ as the slope coefficient of the regression e_t on e_{t-1}. Therefore, to obtain the DW statistic, value in cell D31, type the label 'DW=' and, in cell E31, type the formula **=2(1-SLOPE(C60:C78,D60:D78))**. This will give a value of DW= 1.61 (in cell E31). At the 1% level of significance the DW critical values are $d_L = 0.86$ and $d_u = 1.27$. [The critical values can be obtained manually from the DW table in most statistics books]. As $d_u = 1.27 < DW = 1.61 < 2$, do not reject the null hypothesis of zero serial correlation. This conclusion is also supported by the residual plot given in Figure 12.15.

Task 9 ***Checking for heteroscedasticity***

Figure 12.16 presents the plot of the residuals against the variable X (this plot is given in the output from the Analysis Tool *Regression* in Task 4 when *estimating a linear regression line*).

As can be seen from Figure 12.16, the residuals are scattered at random within a rectangular band and there is no evidence of heteroscedasticity. That is, the errors appear to be homoscedastic, as expected.

Task 10 ***Graphing the actual and predicted values***

The actual values of Y are in cells C1 to C21 and predicted values of Y (Yhat) are in cells B59 to B78. To draw the plot of the actual values (in cells C1 to

Figure 12.16 Graph of residual against X

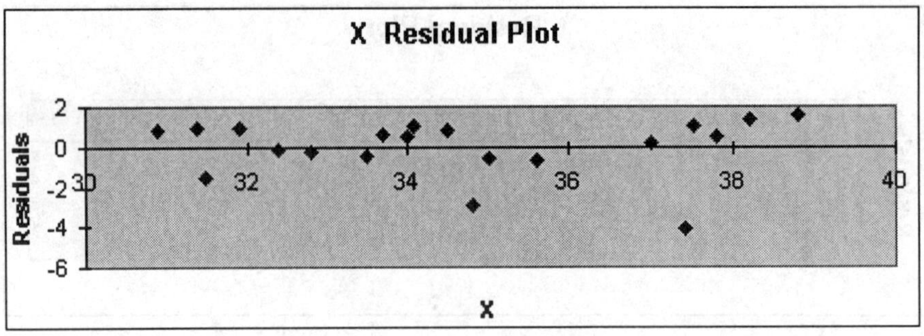

C21) against the predicted values (in cells B59 to B78), we use the 5 steps of ChartWizard. In step 1, type cell references as **C1:C21,B59:B78**. To assess the predictability of the model, we superimpose an estimated linear trend line with zero intercept to show the relationship between Y and Yhat. If the slope of this line is close to one, then that can be considered as the line through the origin at an angle 45°. To draw the linear trend line, we follow the same procedure as in Task 9 of Example 12.1 (Section 12.2). The output is given in Figure 12.17.

Figure 12.17 Graph of predicted against actual sales

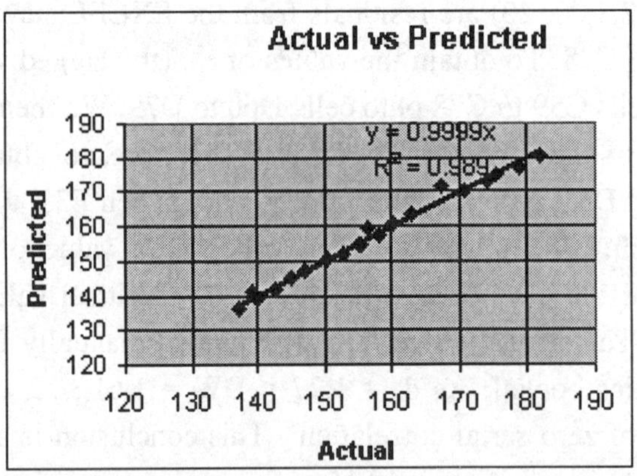

As can be seen, the estimated linear regression line between Y and Yhat has slope = .9999, which is approximately equal to 1, indicating that the points on the plot lie very close to the 45° line through the origin. This shows that the predictability of the model is excellent.

Task 11 ***Forecasting the dependent variable***

To forecast the total sales of all chain stores for the next 4 quarters, in cell C23 type the title 'FORECAST'. In cells A24 to A27, type the values for t **(=21,22,23,24)** and in cells B24 to B27, type the four given values for X **(38.5, 38.8, 39.3** and **38.8)** In cell C24, type the formula **=$B$51+$B$52*A24+$B$53*B24**. Then copy this formula onto cells C25 to C27. The forecast values can be seen in cells C24 to C27.

Task 12 ***Including quarterly dummy variables in multiple regression model***

First copy cells A1 to B21 onto cells E1 to F21. Then, set up the three dummy variables in cells G1 to G21 (Q1), H1 to H21 (Q2) and I1 to I21 (Q3), respectively. [To estimate a multiple regression model in EXCEL, it is a requirement that all the X variables are set-up in a block of columns.] Now, as in Task 4, re-run the regression with Y variable range as cells C1 to C21 and X variables range as E1 to I21. The results can be seen in Figure 12.18. As can be seen from the *p*-values (for testing the significance of the individual coefficients) given in the summary output, none of the dummy variables, Q1, Q2 and Q3, is significant.

Task 13 Save your workbook as '**A:EX122.XLS**' and print the file.

Figure 12.18 Summary multiple regression with dummy variables output

	A	B	C	D	E	F	G	H	I
1	t	x	y		t	x	Q1	Q2	Q3
2	1	30.9	137.3		1	30.9	1	0	0
3	2	31.4	140		2	31.4	0	1	0
4	3	31.9	142.7		3	31.9	0	0	1
5	4	31.5	139.4		4	31.5	0	0	0
6	5	32.4	145		5	32.4	1	0	0
7	6	32.8	147.1		6	32.8	0	1	0
8	7	33.5	150.3		7	33.5	0	0	1
9	8	33.7	152.8		8	33.7	0	0	0
10	9	34.1	155.5		9	34.1	1	0	0
11	10	34	155.3		10	34	0	1	0
12	11	34.5	158.3		11	34.5	0	0	1
13	12	34.8	156.4		12	34.8	0	0	0
14	13	35	160.2		13	35	1	0	0
15	14	35.6	163.1		14	35.6	0	1	0
16	15	37.4	167.3		15	37.4	0	0	1
17	16	37	170.7		16	37	0	0	0
18	17	37.5	174.2		17	37.5	1	0	0
19	18	37.8	175.6		18	37.8	0	1	0
20	19	38.2	178.7		19	38.2	0	0	1
21	20	38.8	181.9		20	38.8	0	0	0

	A	B	C
23			Forecast
24	21	38.5	179.882645
25	22	38.8	181.7490173
26	23	39.3	184.3747518
27	24	38.8	183.2036753

	A	B	C	D	E	F	G
85	SUMMARY OUTPUT						
86							
87	*Regression Statistics*						
88	Multiple R	0.994886					
89	R Square	0.989797					
90	Adjusted R Square	0.986154					
91	Standard Error	1.61298					
92	Observations	20					
93							
94	ANOVA						
95		*df*	*SS*	*MS*	*F*	*gnificance F*	
96	Regression	5	3533.6	706.723	271.638	2E-13	
97	Residual	14	36.424	2.6017			
98	Total	19	3570				
99							
100		*Coefficient*	*andard Er*	*t Stat*	*P-value*	*ower 95%*	*pper 95%*
101	Intercept	13.5308	32.686	0.41396	0.68517	-56.573	83.635
102	t	0.680093	0.4491	1.51451	0.15214	-0.283	1.6432
103	x	3.940503	1.0809	3.64572	0.00265	1.6223	6.2587
104	Q1	0.890073	1.0395	0.85622	0.40629	-1.3395	3.1197
105	Q2	0.650209	1.0283	0.63234	0.53735	-1.5552	2.8556
106	Q3	0.136523	1.0904	0.12521	0.90214	-2.2021	2.4751

12.5 Exercises to Chapter 12

Exercise 12.1

(Ref: Based on Supplementary Exercise 17.68, Australian Business Statistics, Selvanathan et al., 1994)

The store manager of Colonial Furniture Ltd has been reviewing quarterly advertising expenditures. TV spot ads in particular caught her eye, because they were the major expenditure item. In order to maximise cost-effectiveness, she would like to get a better idea of the relationship between the TV spot advertising she sponsors and the number of people (adults) who visit her store because of them. To this end, she has compiled the data contained in the following table.

Task 1 Enter the data in cells A1 to B17 in two columns (X in column A and Y in column B) with column labels 'X' and 'Y' in row 1.

Task 2 Draw a scatter plot of number of people coming to the store (Y) against the number of TV ads run (X).

Task 3 Measure the strength of the linear relationship between the number of people coming to the store (Y) and the number of TV ads run (X).

Task 4 Estimate a linear relationship between the two variables in the form

$$Y = \beta_0 + \beta_1 X + \varepsilon.$$

Obtain the residuals and fitted values. Obtain the plot of Y and predicted value of Y against X on the same graph.

Task 5 Draw the estimated regression line obtained in Task 4 on the same plot from Task 2.

Quarterly number of TV ads run x	Quarterly number of people y	
7	42	
5	32	
1	10	
8	40	
10	61	
2	8	
6	35	
7	39	
8	48	
9	51	
5	30	
7	45	
8	41	
2	7	
6	37	
5	33	
$\sum x = 96$	$\sum y = 559$	$\sum xy = 3930$
$\sum x^2 = 676$	$\sum y^2 = 23037$	

Task 6 What is the value of the coefficient of determination R^2? What is the standard error of estimate s_e? Does it appear that a simple linear regression model is reasonable?

Task 7 Calculate the 95% confidence interval for the slope coefficient β_1.

Task 8 Is the relationship significant at the 5 percent level of significance?

Task 9 Plot the actual values against fitted values and comment on the predictability of your estimated model.

Task 10 Forecast the number of people coming to the store if the number of TV ads run is 6.

Task 11 Plot the residual values against X and comment on the issue of heteroscedasticity.

Task 12 Save your file as 'A:STORE.XLS' on your disk and print the file.

Exercise 12.2

(Ref: Based on Case 18.1: Demand for beer in Australia, <u>Australian Business Statistics</u>, Selvanathan et al., 1994)

For many reasons, the consumption of alcohol continues to attract a lot of attention in the media. On the one hand, the alcohol industry is always interested to know how the market share of its product changes; and looking for ways to increase the demand for alcohol. On the other hand, the government, the health profession and social workers are searching for ways to reduce alcohol consumption by introducing control mechanisms such as increasing alcohol taxes, banning alcohol advertising, using random breath testing, etc.

The following table presents the data for per capita consumption of beer, prices of beer, wine and spirits and per capita income for Australia during the period 1975-91. Using all variables in logarithmic form, analyse the demand for beer in Australia.

Year	Per capita beer consumption Y (in litres)	Per capita income X_1 ($)	Price of beer X_2 ($/litre)	Price of wine X_3 ($/litre)	Price of spirits X_4 ($/litre)
1975	137.39	3 170	0.93	2.71	10.36
1976	135.01	3 567	1.02	2.93	10.77
1977	134.80	3 911	1.07	3.11	11.11
1978	130.81	4 359	1.24	3.26	14.87
1979	132.29	4 858	1.34	3.37	16.51
1980	129.34	5 426	1.46	3.60	17.75
1981	128.63	6 081	1.61	3.83	18.52
1982	121.68	6 814	1.84	4.16	20.08
1983	117.76	7 418	2.01	4.47	21.57
1984	114.46	8 058	2.18	4.78	23.01
1985	115.47	8 863	2.37	4.91	25.22
1986	111.30	9 697	2.61	5.31	27.85
1987	110.80	10 632	2.85	5.64	30.37
1988	113.10	11 588	2.90	6.08	32.99
1989	111.60	12 701	3.08	6.34	36.11
1990	108.20	13 219	3.34	6.39	39.33
1991	101.90	13 685	3.45	6.57	42.33

Source: Selvanathan 1993, 'Alcohol consumption in Australia', *Economics of Drug Abuse*, edited by David Collins and Helen Lapsby, National Campaign Against Druce Abuse Monograph, AGPS, Canberra.

Task 1 Enter the data in cells A1 to E18 in five columns (X_1, X_2, X_3 and X_4 in columns A,B, C and D, respectively, and Y in column E) with column headings 'X1', 'X2,', 'X3', 'X4' and 'Y' in cells A1, B1, C1, D1 and E1.

Task 2 Draw a scatter plot of Y against X_1. Comment on the plot.

Task 3 Convert the variables X_1, X_2, X_3, X_4 and Y to logarithmic form. That is, obtain the variables $x_1 = \ln X_1$, $x_2 = \ln X_2$, $x_3 = \ln X_3$, $x_4 = \ln X_4$ and $y = \ln Y$ in cells F1 to J18, with column headings 'x1 = ln X1', 'x2 = ln X2', 'x3 = ln X3', 'x4 = ln X4' and 'y = ln Y' in cells F1-J1.

Task 4 Measure the strength of the linear relationship between y and x_1.

Task 5 Estimate a multiple linear relationship between y, x_1, x_2, x_3 and x_4 in the form

$$y = \beta_o + \beta_1 x_1 + \beta_2 x_2 + \beta_3 x_3 + \beta_4 x_4 + \varepsilon,$$

where all the variables are in logarithmic form. Obtain the residuals, a residual plot and a line fit plot.

Task 6 Construct a 95% confidence interval estimate for the coefficient β_1.

Task 7 Are the estimated coefficients individually significant at the 5 percent level of significance?

Task 8 What is the value of the coefficient of determination R^2? What is the standard error of estimate s_e? Does it appear that the above model is reasonable? Use a test to assess the overall utility of the model.

Task 9 From the plot of the residuals, explain whether you find evidence of positive autocorrelation. Conduct a formal test for positive autocorrelation using Durbin-Watson test at significance level .01. Comment on whether the residual plot results are in accordance with the DW test results?

Task 10 Do you suspect heteroscedasticity of errors? Use the plot of the residuals against the X_1 to confirm your observation.

Task 11 Plot the actual values against the fitted values and comment on the predictability of your estimated model.

Task 12 Forecast the beer consumption for 1992 if the per capita income is $14,100, prices of beer, wine and spirits for 1992 are $3.55, $6.75, and $45.50, respectively. [Note that you need to convert the data to logarithmic form before using the estimated regression line for forecasting].

Task 13 Save your EXCEL workbook as 'A:BEER.XLS' on your disk and print the file.

Chapter 13

Index Numbers

A numerical measure that measures the changes in a time series over time is called an **Index Number**. Some of the well known such index numbers are the *Consumer Price Index* (CPI), which is used to measure the overall price movements of consumer purchases, and the *Australian All Ordinaries Index*, which is used to measure the general price movements of various stocks in the Australian Stock Exchange.

In practice, various approaches are used to construct index numbers. Below, we look at the calculation of price indices using 5 different approaches. Even though, we construct only price indices here, the same procedures can be used to calculate indices for any other variables (for example quantity indices). We also see how these indices are used to make comparison of movements of another time series by taking out, for example, the price effects.

Even though many applications of index numbers are for time-series, index numbers are also commonly used to make comparisons across different regions/countries, where regions/countries play the role of time.

13.1 Methods of calculating index numbers

Let us consider the prices and quantities purchased of n goods at two different time periods, base period 0 and current period 1. Let p_{i0} and q_{i0} be the price and quantity of the ith good in some base period 0; and p_{i1} and q_{i1}, be the price and quantity of the ith good in the current period 1, for i=1,2, …, n.

	Period 0		Period 1	
Good	Price	Quantity	Price	Quantity
1	p_{10}	q_{10}	p_{11}	q_{11}
2	p_{20}	q_{20}	p_{21}	q_{21}
3	p_{30}	q_{30}	p_{31}	q_{31}
.	...	...	...	...
.	...	...	...	...
n	p_{n0}	q_{n0}	p_{n1}	q_{n1}

Simple aggregate price index

A *simple aggregate price index* is the ratio of the sum of prices of all commodities in the current period 1 over the sum in the base period 0, multiplied by 100. That is,

$$I_{1,0} = \frac{\sum_{i=1}^{n} p_{i1}}{\sum_{i=1}^{n} p_{i0}} \times 100$$

Weighted aggregate price index

A *weighted aggregate price index* is the ratio of the weighted sum of prices of all commodities in the current period 1 over the weighted sum in the base period 0 (the weights being the corresponding quantity proportions), multiplied by 100. That is,

$$I_{1,0} = \frac{\sum_{i=1}^{n} w_{i1} p_{i1}}{\sum_{i=1}^{n} w_{i0} p_{i0}} \times 100$$

where the weights are given by

$$w_{i0} = \frac{q_{i0}}{\sum_{i=1}^{n} q_{i0}}$$

and

$$w_{i1} = \frac{q_{i1}}{\sum_{i=1}^{n} q_{i1}}$$

Note that $\sum_{i=1}^{n} w_{i0} = 1$ and $\sum_{i=1}^{n} w_{i1} = 1$.

Laspeyres price index

The *Laspeyres price index* is the ratio of the weighted sum of prices of all commodities in the current period 1 over the weighted sum in the base period 0 (the weights being the period 0 quantities), multiplied by 100. That is,

$$L_{1,0} = \frac{\sum_{i=1}^{n} p_{i1} q_{i0}}{\sum_{i=1}^{n} p_{i0} q_{i0}} \times 100$$

Paasche price index

The *Paasche price index* is the ratio of the weighted sum of prices of all commodities in the current period 1 over the weighted sum in the base period 0 (the weights being the period 1 quantities), multiplied by 100. That is,

$$P_{1,0} = \frac{\sum_{i=1}^{n} p_{i1} q_{i1}}{\sum_{i=1}^{n} p_{i0} q_{i1}} \times 100$$

Fisher price index

The *Fisher price index* is the geometric mean of Laspeyres and Paasche indices. That is,

$$F_{1,0} = \sqrt{L_{1,0} \times P_{1,0}}$$

Aim: At the completion of the examples in this chapter you should be able to use EXCEL to calculate various indices such as the simple aggregate index, weighted aggregate index, Laspeyres index, Paasche index and Fisher index; and to deflate a time series using another index series.

13.2 Example 13.1: Calculating price indexes

(Ref: Based on Exercise 22.27, p.744, Australian Business Statistics, Selvanathan et al, 1994)

Apparent consumption of meat consumption and average prices in Sydney in 1984, 1986 and 1988 are summarised in the following table.

	Price/kg ($)			Per capita annual consumption(kg)		
	1984	1986	1988	1984	1986	1988
Beef	7.57	8.02	8.99	40.0	37.5	40.1
Lamb	3.86	3.83	4.61	17.0	15.0	14.9
Pork	4.39	4.90	5.16	16.4	16.8	17.5
Chicken	2.50	2.72	2.59	21.8	23.0	24.3

Source: *Australian Bureau of Statistics, Average retail prices of selected items and apparent per capita consumption*, Cat. no. 4309, ABS, Canberra.

Using EXCEL compute the following price indices to measure the meat price increases since 1984.

Task 1 Compute the simple aggregate price index.

Task 2 Compute the weighted aggregate price index.

Task 3 Compute the Laspeyres price index.

Task 4 Compute the Paasche price index.

Task 5 Compute the Fisher price index.

Task 6 Save your workbook as 'A:EX131.XLS' on your disk and print the file.

Solution to Example 13.1

In cells B1 to G1, type the column headings 'p84', 'p86', 'p88', 'q84', 'q86' and 'q88' and centre them. In cells A2 to A5, type the labels 'Beef', 'Lamb', 'Pork' and 'Chicken'. Enter the given prices and consumption in cells B2 to G5.

Task 1 ***Simple aggregate price index***

In cell A9, type the title 'Simple Aggregate Price Index'. In cell B8, C8 and D8, type the column headings 'I84,84', 'I86,84' and 'I88,84' and centre them. In cells B9 to B13, type **100** in every cell. To calculate the simple aggregate price index, we use the formula given in Section 13.1. In cell C9 and D9, type the formulas **=(SUM(C2:C5)/SUM(B2:B5))*100** and **=(SUM(D2:D5)/SUM(B2:B5))*100.**

Task 2 ***Weighted aggregate price index***

First we calculate the weights defined in Section 13.1. In cells H1 to J1, type the column headings 'W84', 'W86' and 'W88' and centre them. In cells E6 to G6, type the formulas **=SUM(E2:E5), =SUM(F2:F5) and =SUM(G2:G5).** To obtain the weights, in cell H2, type the formula **=E2/$E$6** and copy this formula onto cells H3 to H5; in cell I2, type the formula **=F2/$F$6** and copy this formula onto cells I3 to I5; and, in cell J2, type the formula **=G2/$G$6** and copy this formula onto cells J3 to J5.

In cell A10, type the title 'Weighted Aggregate Price Index'. To calculate the weighted aggregate price index, we use the formula given in Section 13.1. Therefore, in cells C10 and D10, type the formulas

=(SUMPRODUCT(C2:C5,I2:I5)/SUMPRODUCT(B2:B5,H2:H5))*100 and **=(SUMPRODUCT(D2:D5,J2:J5)/SUMPRODUCT(B2:B5,H2:H5))*100.**

Task 3 ***Laspeyres price index***

In cell A11, type the label 'Laspeyres Price Index'. To calculate the Laspeyres price index, we use the formula given in Section 13.1. Therefore, type the formulas **=(SUMPRODUCT(C2:C5,E2:E5)/SUMPRODUCT(B2:B5,E2:E5))*100** and **=(SUMPRODUCT(D2:D5,E2:E5)/SUMPRODUCT(B2:B5,E2:E5))*100** in cells C11 and D11.

Task 4 ***Paasche price index***

In cell A12, type the label 'Paasche Price Index'. To calculate the Paasche price index, we use the formula given in Section 13.1. Therefore, type the formulas **=(SUMPRODUCT(C2:C5,F2:F5)/SUMPRODUCT(B2:B5,F2:F5))*100** and **=(SUMPRODUCT(D2:D5,G2:G5) /SUMPRODUCT (B2:B5,G2:G5))*100** in cells C12 and D12.

Task 5 ***Fisher price index***

In cell A13, type the label 'Fisher Price Index'. To calculate the Fisher price index, we use the formula given in Section 13.1. Therefore, in cell C13 and D13, type the formulas **=SQRT(C11*C12)** and **=SQRT(D11*D12).**

Task 6 Save your workbook as **A:EX131.XLS** on your disk and print the file. Figure 13.1 presents the output.

Figure 13.1 Simple aggregate, weighted aggregate, Laspeyres, Paasche and Fisher price indexes

	A	B	C	D	E	F	G	H	I	J
1		p84	p86	p88	q84	q86	q88	w84	w86	w88
2	Beef	7.57	8.02	8.99	40	37.5	40.1	0.4201681	0.40628386	0.4142562
3	Lamb	3.86	3.83	4.61	17	15	14.9	0.1785714	0.16251354	0.15392562
4	Pork	4.39	4.9	5.16	16.4	16.8	17.5	0.1722689	0.18201517	0.18078512
5	Chicken	2.5	2.72	2.59	21.8	23	24.3	0.2289916	0.24918743	0.25103306
6					95.2	92.3	96.8	1	1	1
7										
8		184,84	186,84	188,84						
9	Simple Aggregate Price Index	100	106.28	116.54						
10	Weighted Aggregate Price Index	100	104.84	115.74						
11	Laspeyres Price Index	100	106.19	117.00						
12	Paasche Price Index	100	106.35	116.80						
13	Fisher Price Index	100	106.27	116.90						

13.3 Example 13.2: Deflating a time series using the CPI

(Ref: Based on Example 22.7, p.738, Australian Business Statistics, Selvanathan et al, 1994)

The Gross Domestic Product (GDP) is often used as a measure of the economic growth of a country. The Australian GDP (in millions of current \$) for the years 1984 to 1990 together with the Consumer Price Index, CPI(1980=100), are shown below.

Year	GDP(Current m$)	CPI(1980=100)
1984	211,915	137.2
1985	235,815	148.7
1986	260,367	162.6
1987	298,335	174.5
1988	340,440	187.3
1989	370,805	202.3
1990	377,114	213.0

Source: *Australian Bureau of Statistics*, Cat. nos. 3101.0 & 5206.0, ABS, Canberra.

Use EXCEL to perform the following tasks.

Task 1 Convert the CPI(1980=100) series to CPI(1985=100) series.

Task 2 Use the CPI(1985=100) figures obtained in Task 1 to deflate the GDP (in millions of current $) to 1985 constant millions of dollars.

Task 3 Save your workbook as 'A:EX132.XLS' on your disk and print the file.

Solution to Example 13.2

In cells A1 to C1, type the column headings 'Year', 'GDP(current m$)' and 'CPI(1980=100)'. In cells A2 to C8, type the given data.

Task 1 ***Converting the CPI series***

In cell D1, type the column heading 'CPI(1985=100)'. To convert the CPI(1980=100) series (in cells C2 to C8) to CPI(1985=100), divide each entry in that column by the 1985 CPI value in cell C3. First in cell D2, type the formula **=(C2/$C$3)*100** and copy this formula onto cells D3 to D8.

Task 2 ***Deflating the GDP series in current prices to constant prices***

In cell E1, type the column heading 'GDP(1985 constant m$)'. To convert the GDP(current m$) series to GDP(1985 constant m$) series, we deflate GDP (current m$) by the CPI(1985=100). In cell E2, type the formula **=(B2/D2)*100** and copy this formula onto cells E3 to E8.

Task 3 Save your workbook as **A:EX132.XLS** on your disk and print the file. Figure 13.2 presents the output.

Figure 13.2 Actual and deflated GDP

	A	B	C	D	E
1	Year	GDP(Current m$)	CPI(1980=100)	CPI(1985=100)	GDP(1985 Constant m$)
2	1984	211915	137.2	92.27	229677.55
3	1985	235815	148.7	100.00	235815.00
4	1986	260367	162.6	109.35	238109.30
5	1987	298335	174.5	117.35	254225.87
6	1988	340440	187.3	125.96	270279.91
7	1989	370805	202.3	136.05	272559.09
8	1990	377114	213	143.24	263271.60

13.4 Exercises to Chapter 13

Exercise 13.1

(Ref: Based on Exercise 22.4, p.302, Australian Business Statistics, Study Guide, Selvanathan et al, 1995)

The following computer components were ordered by a large industrial concern in years 1987 and 1990.

	Price ($)		Quantity	
	1987	**1990**	**1987**	**1990**
Disk Drives	339.0	279.0	30	70
Monitors	379.0	449.0	100	50
Printers	669.0	379.0	40	50
PC's	1499.0	899.0	100	40

Using EXCEL, compute the following price indices to measure the computer component price increases since 1987.

Task 1 Compute the simple aggregate price index.
Task 2 Compute the weighted aggregate price index.
Task 3 Compute the Laspeyres price index.
Task 4 Compute the Paasche price index.
Task 5 Compute the Fisher price index.
Task 6 Save your workbook as 'A:COMPO.XLS' on your disk and print the file.

Exercise 13.2

(Ref: Based on Ex. 22.20, p.743, Australian Business Statistics, Selvanathan et al, 1994)

Is real estate a good investment? In particular, will the price of a house keep up with inflation? To answer these questions, the median sales price of privately owned houses in Sydney and CPI (1980=100) for Sydney were recorded for 1980 to 1988. These data are shown below.

Year	Median House Price (Current $)	CPI (1980=100)
1980	64,700	100.0
1981	79,600	110.2
1982	81,900	123.4
1983	80,400	130.9
1984	84,300	136.0
1985	89,700	147.5
1986	98,000	161.4
1987	109,800	173.7
1988	142,000	187.7

Use EXCEL to perform the following tasks.

Task 1 Convert CPI(1980=100) series to CPI(1985=100) series.

Task 2 Use the CPI(1985=100) figures obtained in Task 1 to deflate the Sydney house prices (in current $) to 1985 constant dollars.

Task 3 Save your workbook as 'A:HOUSE.XLS' on your disk and print the file.

Appendix

Statistical Functions in EXCEL

The following is a list of statistical functions available in EXCEL and their brief description. More details can be obtained using the Help facility in EXCEL.

AVEDEV	Returns the average of the absolute deviations of data points from their mean
AVERAGE	Returns the average of its arguments
BETADIST	Returns the cumulative beta probability density function
BETAINV	Returns the inverse of the cumulative beta probability density function
BINOMDIST	Returns the individual term binomial distribution probability
CHIDIST	Returns the one-tailed probability of the chi-squared distribution
CHIINV	Returns the inverse of the one-tailed probability of the chi-squared distribution
CHITEST	Returns the test for independence
CONFIDENCE	Returns the confidence interval for a population mean
CORREL	Returns the correlation coefficient between two data sets
COUNT	Counts how many numbers are in the list of arguments
COUNTA	Counts how many values are in the list of arguments
COVAR	Returns covariance, the average of the products of paired deviations
CRITBINOM	Returns the smallest value for which the cumulative binomial distribution is less than or equal to a criterion value
DEVSQ	Returns the sum of squares of deviations
EXPONDIST	Returns the exponential distribution
FDIST	Returns the F probability distribution
FINV	Returns the inverse of the F probability distribution
FISHER	Returns the Fisher transformation
FISHERINV	Returns the inverse of the Fisher transformation
FORECAST	Returns a value along a linear trend
FREQUENCY	Returns a frequency distribution as a vertical array
FTEST	Returns the result of an F-test
GAMMADIST	Returns the gamma distribution
GAMMAINV	Returns the inverse of the gamma cumulative distribution

GAMMALN	Returns the natural logarithm of the gamma function, G(x)
GEOMEAN	Returns the geometric mean
GROWTH	Returns values along an exponential trend
HARMEAN	Returns the harmonic mean
HYPGEOMDIST	Returns the hypergeometric distribution
INTERCEPT	Returns the intercept of the linear regression line
KURT	Returns the kurtosis of a data set
LARGE	Returns the k-th largest value in a data set
LINEST	Returns the parameters of a linear trend
LOGEST	Returns the parameters of an exponential trend
LOGINV	Returns the inverse of the lognormal distribution
LOGNORMDIST	Returns the cumulative lognormal distribution
MAX	Returns the maximum value in a list of arguments
MEDIAN	Returns the median of the given numbers
MIN	Returns the minimum value in a list of arguments
MODE	Returns the most common value in a data set
NEGBINOMDIST	Returns the negative binomial distribution
NORMDIST	Returns the normal cumulative distribution
NORMINV	Returns the inverse of the normal cumulative distribution
NORMSDIST	Returns the standard normal cumulative distribution
NORMSINV	Returns the inverse of the standard normal cumulative distribution
PEARSON	Returns the Pearson product moment correlation coefficient
PERCENTILE	Returns the k-th percentile of values in a range
PERCENTRANK	Returns the percentage rank of a value in a data set
PERMUT	Returns the number of permutations for a given number of objects
POISSON	Returns the Poisson distribution
PROB	Returns the probability that values in a range are between two limits
QUARTILE	Returns the quartile of a data set
RANK	Returns the rank of a number in a list of numbers
RSQ	Returns the square of the Pearson product moment correlation coefficient
SKEW	Returns the skewness of a distribution
SLOPE	Returns the slope of the linear regression line
SMALL	Returns the k-th smallest value in a data set
STANDARDIZE	Returns a normalised value
STDEV	Estimates standard deviation based on a sample
STDEVP	Calculates standard deviation based on the entire population
STEYX	Returns the standard error of the predicted y-value for each x in the regression
TDIST	Returns the Student's t-distribution
TINV	Returns the inverse of the Student's t-distribution
TREND	Returns values along a linear trend
TRIMMEAN	Returns the mean of the interior of a data set
TTEST	Returns the probability associated with a Student's t-Test
VAR	Estimates variance based on a sample
VARP	Calculates variance based on the entire population
WEIBULL	Returns the Weibull distribution
ZTEST	Returns the two-tailed P-value of a z-test

Index